Advance praise for **GREENING MEDIA EDUCATION**

"With his new book, Antonio López has established himself as a leading thinker in the emerging field of ecomedia literacy. His book is a must read for anyone concerned about the transformation that must be made in media education if it is contribute to an ecologically sustainable future. His insightful use of an ecological conceptual framework provides the field with the needed sense of direction that moves beyond the misconceptions of the past."
—Chet Bowers, Author, *The False Promises of the Digital Revolution* and *In the Grip of the Past: Educational Reforms that Address What Should be Changed and What Should be Conserved*

"As an activist-scholar working in the 'strange' (after Maxine Greene) intersectional curricular spaces of media and ecological literacy, Antonio López has become a crucial leader who is assisting both academics and citizens with the reimagining of sustainability education for a present moment that is now co-constituted by planetary ecocrisis and global media culture. To my mind, *Greening Media Education* represents his most critically exact, compendious, and powerful research to date on this emergent paradigm of democratic ecopedagogy and politics. This is a necessary and very important work."
—Richard Kahn, Core Faculty in Education, Antioch University Los Angeles; Author, *Critical Pedagogy, Ecoliteracy and Planetary Crisis: The Ecopedagogy Movement*

"*Greening Media Education* draws the best from the media studies cannon and takes on unelected educational powers using a well-calibrated eco-ethical compass. The author's deep environmental commitments provide a much-needed green perspective on the practices of contemporary media literacy."
—Richard Maxwell, Chair and Professor of Department of Media Studies, Queens College-CUNY; Author, *Greening the Media*

"Antonio López has been at the forefront of activist media education research in recent years. In *Greening Media Education* he weaves together the multiple strands of his praxis to provide a compelling vision of a more situated, sustainable and resistant pedagogy. In the era of the 'creative economy' and neoliberal instrumentalism, it may be crazily aspirational— but Lopez writes from the heart with profound optimism about media education for social justice."
—Julian McDougall, Associate Professor in Media and Education and Director of the Centre for Excellence in Media Practice, Bournemouth University, United Kingdom

More advance praise for **GREENING MEDIA EDUCATION**

"This book fills an important gap in media studies and media education. It should be read by everyone concerned about the future of our planet and the way we have come to understand it. This is a significant and timely book combining theoretical insight with practical and empirical understanding. Sustainability and media literacy are not two separate entities but intimately related. Antonio López shows why this is the case and why media educators need to urgently integrate the two. [This] pioneering work in media education and environmental sustainability issues is both timely and significant. It is a major contribution that deserves to be read and discussed widely."

—John Blewitt, Aston University, Birmingham, United Kingdom;
Author, *Understanding Sustainable Development*

"With *Greening Media Education* Antonio López provides an essential framework for K-12 educators who seek to integrate critical thinking media literacy into sustainability education. This work will help teachers and students support the flowering of a media literacy ecosystem where social justice and environmental stewardship are joined at the root."

—Sox Sperry, Author, *Media Constructions of Sustainability: Food, Water and Agriculture*

"Antonio López provides much-needed critique of the current state of media literacy. In doing so, he maps out a new direction for us to take, prioritizing sustainability and social justice. His multicultural perspective provides an intersectional framework, creating critically nuanced solutions that we can all learn from."

—Andrea Quijada, Executive Director, Media Literacy Project

# GREENING MEDIA EDUCATION

CRITICAL ISSUES FOR LEARNING AND TEACHING

Shirley R. Steinberg and Pepi Leistyna
*General Editors*

Vol. 13

The Minding the Media series is part of both
the Peter Lang Education list and the Media and Communication list.
Every volume is peer reviewed and meets
the highest quality standards for content and production.

PETER LANG
New York • Bern • Frankfurt • Berlin
Brussels • Vienna • Oxford • Warsaw

# ANTONIO LÓPEZ

# GREENING MEDIA EDUCATION

Bridging Media Literacy with
Green Cultural Citizenship

PETER LANG
New York • Bern • Frankfurt • Berlin
Brussels • Vienna • Oxford • Warsaw

Library of Congress Cataloging-in-Publication Data
López, Antonio.
Greening media education: bridging media literacy
with green cultural citizenship / Antonio López.
pages cm. — (Minding the media: critical issues for learning and teaching; vol. 13)
Includes index.
1. Media literacy. 2. Environmental education. I. Title.
P96.M4L668   302.23—dc23   2014009160
ISBN 978-1-4331-2591-1 (hardcover)
ISBN 978-1-4331-2590-4 (paperback)
ISBN 978-1-4539-1350-5 (e-book)
ISSN 2151-2949

Bibliographic information published by **Die Deutsche Nationalbibliothek**.
**Die Deutsche Nationalbibliothek** lists this publication in the "Deutsche
Nationalbibliografie"; detailed bibliographic data are available
on the Internet at http://dnb.d-nb.de/.

Cover photo by Antonio López

© 2014 Antonio López
Peter Lang Publishing, Inc., New York
29 Broadway, 18th floor, New York, NY 10006
www.peterlang.com

All rights reserved.
Reprint or reproduction, even partially, in all forms such as microfilm,
xerography, microfiche, microcard, and offset strictly prohibited.

"We have one communications ecosystem and our job is to make it work for everyone."

Michael Copps, FCC Commissioner 2001-2011

Dedicated to all those educating for the seventh generation.

# Contents

Acknowledgments ............................................................................... ix
Introduction: Defamiliarizing Media Literacy ............................................. 1
Chapter One: Media, Environment, and Education ................................. 21
Chapter Two: Metaphors as Meaning Design ......................................... 45
Chapter Three: A Field Walk Through the Media Ecosystem ................. 67
Chapter Four: Mapping the Media Literacy Ecosystem .......................... 87
Chapter Five: The Media Literacy Ecosystem's Dominant Paradigm ..... 113
Chapter Six: Ecomedia Literacy ............................................................. 133
Chapter Seven: Media as Sustainability Education ............................... 161
Bibliography ........................................................................................ 175
Index .................................................................................................... 195

# Acknowledgments

This book is the result of more than a dozen years of exploration, experimentation, and research. However, the bulk of the research, in particular for the section on the media literacy ecosystem, came out of my dissertation work at Prescott College. (This book departs from the dissertation in many ways, in particular by eliminating the methods section and many details of the research. Because of the vicissitudes of digital publishing formats, the publisher also requested the removal of all tables. Much of the data I used was nicely visualized in tables, so if you wish to review the full dissertation, it is posted on this book's web page: www.greenmediaed.com). Therefore, it is necessary to acknowledge those who have contributed in some way to the process and creation of this research. First and foremost, I want to thank my committee chair, Pramod Parajuli, who has the magical ability to ask the right questions and to situate research problems in their proper context. I am grateful for the input and support of my primary committee members, Tema Milstein and John Blewitt, for their belief in my research and for taking time from their busy professional activities to work with me. Thanks also to Bryan Alexander who came in at a crucial moment to offer his expertise and insights. The path to my dissertation involved several mentors whose thoughts, ideas, and inspiration are part of the final product. In particular, I offer thanks to Chet Bowers, Chellis Glendinning, Kathleen Tyner, Adrian J. Ivakhiv, Jennifer Thom, and Rich Lewis. In addition I would like to thank the following for assistance, input, and feedback: Renee Hobbs, Cyndy Scheibe, Frank Baker, Andrea Quijada, Julian McDougall, Ryan Goble, Sox Sperry, and Steve Goodman. Special thanks for this book's series editor, Shirley Steinberg, for green-lighting publication. Finally, hugs and kisses for my dear partner, Cristina Guardata, who endured this writing process with great strength. Thanks for holding the fort!

During the five years of research and writing that went into this book, along the way some of the written materials became the basis for several published pieces. As a result, a few sections in this dissertation also appear in the following published pieces under my name: "Defusing the Cannon/Canon: An Organic Media Approach to Environmental Communication" (2010);

"Greening a Digital Media Course" (2011a); "Greening Media Education" (2011b); "Practicing Sustainable Youth Media" (2011c); and "Greening a Digital Media Course: A Field Report" (2013).

INTRODUCTION

# Defamiliarizing Media Literacy

David Buckingham (2007) posits that media literacy is the outcome of media engagement, but media education is what shapes practice. In the vernacular use of educators and policymakers in North America, "media literacy" is the more common term for the formal and informal process of teaching with or about media. Media literacy education concerns the pedagogy of media literacy, and media literacy educators are the practitioners who are involved with shaping, promoting, and defining the goals of media literacy.

For more than a dozen years I have been a media literacy educator. I am also an environmentalist deeply committed to education for sustainability. As I define it in this book, sustainability education encourages *whole-systems thinking* that is ecological and participatory (Sterling, 2004, p. 11). Sustainability education promotes *green cultural citizenship*, which means embodying sustainable behaviors and cultural practices that shape and promote ecological values within the interconnected realms of society, economy, and environment. In my everyday practice I try to unite perspectives from the fields of media and sustainability education, but having a foot in both worlds has been a struggle. In the process of developing a middle way I have encountered resistance from both educational cultures. Though media literacy advocates often sympathize with environmental issues, the general practice of media literacy marginalizes ecological perspectives. Likewise, there are many in the field of environmental education who believe media and technology are anti-nature (Bowers, 2000; Traina, 1995). Mediating these differences to find common ground has become my life work and is the purpose of this book. As such, in this book I take a deep dive into my community of practice to analyze why this disconnection exists. By mapping the field, I propose a potential solution.

To contextualize my perspective, I would like to share two pivotal experiences, one at a popular congregation of environmentalists and another at an international media literacy conference in Europe. The first occurred in 2003 at the Bioneers conference in San Rafael, California, which is a gathering of "social and scientific innovators from all walks of life and disciplines who have peered deep into the heart of living systems to understand how nature operates, and to mimic 'nature's operating instructions' to serve human ends without harming the web of life" (Bioneers, 2013). This annual event features a variety of visionary thinkers and activists who are developing solutions for a safer and healthier world. I attended the conference with a group of Latinos and Native Americans from New Mexico with the support of a grant from the Pond Foundation. We were invited to encourage cultural diversity at the conference. Being of mixed cultural heritage (Latino/Euro-American), I am accustomed to playing the role of a "bridger"—one who mediates between different social groups and worldviews. At the time of the conference I was a media literacy educator working primarily in Native American and Latino communities. During that time I was balancing my role as an educator of critical thinking tools, media technologies, popular culture, and digital storytelling with the cultural reality of indigenous and land-based youths in rural New Mexico. The conference was an exciting opportunity to see how the various perspectives I was negotiating would intersect with leading sustainability models.

I attended a session led by global justice activist and anti-technology crusader Jerry Mander (1991, 2002). The panel was organized around themes developed in a book he co-edited, *The Case Against the Global Economy: And for a Turn Toward the Local* (Mander & Goldsmith, 1996), which focuses on the rise of globalization and the danger it poses for traditional cultures, economies, and environments. These were relevant issues for the communities in which I was working, for many of them were experiencing the consequence of the privatization of local resources, such as water, and the impact on their ecosystems of military research at Los Alamos National Laboratory. Throughout the presentation I kept seeing the connection between media and the ideology of globalization, helping me realize that media literacy could be used as a tool for students to understand and debate dominant economic discourses. Furthermore, a seed was planted that media literacy could be an invaluable tool for sustainability education. After the panel I approached Mander and asked if he was willing to meet with me to discuss the connection I was making between media education and global justice activism. He graciously accepted and later that day he joined me at a community table near the conference's main bookstore. After asking him to sign my personal copy of *Four Arguments for the Elimination of Television* (2002), I then presented my concept. I suggested that an excellent way to introduce the complexity of globalization to young people is through media literacy. I suggested that using media texts as probes and *objects-to-think-with* could promote discussion and dialogue about globalization and social justice debates. Furthermore, I proposed that media literacy could be a way to introduce sustainability to students, just as it had been used by public health advocates to teach young people about the hazards of smoking and drinking alcohol. I suggested that we could work together to develop this project, but he answered me rather unexpectedly. He told me that he thought media literacy was a good idea, but he was against it. When I asked why, he replied that it is because media literacy "makes media more interesting."

And that was the end of our discussion. But the encounter prompted an inner dialogue that continues in this book. Education should interest learners in the world around them, a world that is highly mediated. Nonetheless, I understand the spirit of Mander's response, which is that potentially media literacy makes media more attractive. Indeed, some research suggests that didactic media literacy produces a boomerang effect by encouraging the opposite behavior it was intended to mitigate (for an overview of the research, see Banerjee & Kubey, 2014). So rather than encourage critical thinking, when practiced in certain ways, media literacy potentially can make media consumption more enticing to students. Moreover, there is a school of thought often identified as Neo-Luddism that views media and technology as desensitizing us from living systems (Bowers, 2000; Ellul, 1964; Glendinning, 1994; Mander, 1991; Mumford, 1970; Sale, 1996). I should have realized that what I proposed went against the kinds of arguments Mander has made in his writings and activism for the past 30 years. Yet, I felt very uncomfortable with his response, for I knew through personal experience that media literacy is very empowering. Nonetheless, the discussion with him prompted me towards this inquiry: When it comes to sustainability education, would media literacy encourage unsustainable cultural practices? Or could it be part of the solution?

Fast-forward nine years later. In November 2012, in the midst of doing my primary research for this book, I participated in the Media & Learning conference in Brussels, Belgium. This international gathering was touted as an opportunity to promote "media wisdom" in Europe, bringing "together practitioners and policy makers who want to contribute to the development of digital and media skills in education and to find new and effective ways to embed media into the learning process" (Media & Learning, 2012). I attended the conference in order to network with other European media educators and to co-present a talk, *Greening Media Education*, with my colleague John Blewitt. The pre-conference negotiations that led to our presentation were fraught with frustration. In our original proposal we requested an hour-long workshop, a format that would allow us to best present our material. Instead, we were placed in a grab-bag session during the last time slot of the conference, a late-afternoon panel that featured two other speakers presenting on completely unrelated topics. Initially, we were granted only 15 minutes for our entire presentation. We argued for additional time and in the end were given 20 minutes. Though our talk was streamed on the web to a wider audience, by the time we presented there were only about a dozen people in the auditorium. When asked if they had any questions, none responded. Unfortunately, I experienced the same lack of audience participation and marginalization at a similar media education conference held the year before in London.

During the Brussels conference I had one particular experience that illustrates the issues raised in this book. The day before my presentation I encountered the head of a major media literacy organization. I mentioned that I was investigating the discourses of media literacy educators to search for a bridge between media literacy and sustainability education. She responded by saying that she did not see any connection. I replied by discussing the recent scientific findings that demonstrate a significant increase of $CO_2$ in the atmosphere, predicting that we might be nearing a tipping point where the opportunity to prevent a global ecological catastrophe would be lost (Barnosky et al., 2012). I proposed that as media literacy educators we had a responsibility to address this problem because within the next ten years all the gadget usage promoted by our education practices would contribute to the doubling of current $CO_2$ emissions generated by the internet, which is already the equivalent of the global airline industry (Cubitt, Hassan, & Volkmer, 2011). She wondered how $CO_2$ emissions could possibly have anything to do with media, and I responded that because all our gadgets are tethered to server farms, our internet usages are mostly powered by coal (Cubitt et al., 2011). She then replied that sustainability was one of many possible issues that could be taken up by media literacy educators, but it did not demand special attention. With that, she excused herself to prepare for her session.

My conversation with Mander in 2003 and the experience at the Brussels conference nine years later bookend a spectrum of views about the relationship between sustainability education **and** media literacy, where sustainability educators view media education antagonistically and for media educators sustainability is seen as an unrelated, irrelevant issue. These incidents were not atypical of the many encounters I have had in the past dozen years. My struggle has been to define the niche where these seemingly opposed views can be reconciled. Consequently, I have felt a desire to identify more clearly the taken-for-granted world in which my peers in the media literacy community operate. Since it is a world in which I am deeply involved, it was important to investigate my practitioner community with a renewed perspective.

Ideally, it should be possible to develop a framework that combines media literacy and ecoliteracy, but as my past experience demonstrates, an ontological difference between the disciplines that inform these educational approaches makes the process difficult. Media literacy is greatly influenced by assumptions formulated within the traditions of media studies and communications, which in turn have been guided by the dominant paradigm of mechanism arising from the Industrial and Scientific Revolutions. Sustainability education challenges the assumptions of this prevailing paradigm. Thus, media literacy and sustainability education discourses (the way practitioners communicate about their respective fields) are substantially different. As noted by Meadows (1991, p. 4),

> Your paradigm is so intrinsic to your mental processes that you are hardly aware of its existence, until you try to communicate with someone with a different paradigm. Listen to an ecologist talk with an economist, a pro-lifer with a pro-choicer, a right-winger with a left-winger. In the difficulties of cross-paradigm discussion, both parties begin to be aware, often uncomfortably, of unspoken, fundamental assumptions they do not share.

For this reason, "It can be an important strategy for people to seek to understand [discourses] better and bring them to overt attention when there are conflicts in communication" (Gee, 2011a, p. 172). Though there may not be an explicit communication conflict between media literacy and sustainability educators, the fact that there are so few resources that combine the two perspectives demonstrates that there is an implicit boundary between them that is taken for granted. Therefore, mapping this boundary is essential for bridging the disciplinary approaches of media literacy and sustainability education.

In order to gain insights into these disciplinary boundaries, this book features an investigation into how media literacy practitioners use metaphors to frame both the role of media education in the world and how it affects green cultural citizenship. This involved analyzing website documents and teacher resources of seven North American media literacy organizations as well as interviewing nine key practitioners within a bounded system I call the *media literacy ecosystem* (a network of practitioners who share a generally agreed-upon worldview about media literacy, see below). Drawing on an ecocritical framework, I analyzed the discourses of the media literacy ecosystem by using multi-site situational analysis, qualitative media analysis, and critical discourse analysis. As a result, this book explores how media literacy practitioners participate in meaning-making systems that reproduce pre-existing environmental ideologies. The findings show that media literacy education is grounded in a *mechanistic* worldview, thereby perpetuating unsustainable cultural practices in education. By problematizing the mechanistic discourses of media literacy education, the aim of this book is to raise awareness and to offer potential solutions for changing the nature of those same discourses. As such, I propose a model of media literacy that incorporates green cultural citizenship, called ecomedia literacy, and outline a path forward so that sustainability becomes a priority for media literacy educators.

During my research I deployed two strategies often used by media literacy educators to analyze media. The first is to defamiliarize the familiar. Because media are all around us and they embed taken-for-granted assumptions about the world, it can be difficult to see them with fresh eyes. Media literacy educators use a variety of tactics to defamiliarize everyday media, including deconstruction and media production, each offering different ways to critically engage with media. The second is to treat media texts as objects-to-think-with to provoke questions and generate new ideas.

I attempted both strategies by starting with the assumption that media literacy involves mediamaking: as educators we design teaching materials, engage in online discussions, produce web pages, edit videos, write books, and scribe articles. I wanted to investigate the media that we use in our daily practice as media literacy educators, but defamiliarize myself with them through discourse analysis to identify commonly used metaphors. To do so, I created visual documentation of the data I collected by using tools such as word clouds. This allowed me to see general patterns and juxtapositions of terms that practitioners use to describe their worldview. But my ultimate tool for defamiliarization was to re-conceptualize the entire field of media education as a "media literacy ecosystem." This strategy resonated with a number of novel approaches that explain social relations, media, technology, and communications by using ecosystem models (Altheide, 1995; Guattari, 2008; Luhmann, 1989; Nardi & O'Day, 2000; Naughton, 2012; Tracy, 2012). By viewing media literacy as a dynamic system of practitioners interacting with other social systems (such as education and media), I was able to understand more clearly how media educators share a certain worldview.

As Greenwood and Levin (2007, p. 69) assert, "The role of theories is to explain how what happened was possible and took place, to lay out possible scenarios for the future, and to give good reasons for the ones that seem to be the probable next outcomes." Because currently there are very few resources or methodologies that combine media literacy education with green cultural citizenship, this study can contribute to an understanding for how media education practice can be greened. Additionally, the study's results will add to the body of knowledge regarding the methodology of sustainability education. Of significance is having a reflective map of the particular meaning design that governs contemporary media literacy practices. Given that root metaphors about media, communication, and learning determine the kinds of questions and problems to be studied, it is significant to situate these metaphors and the kind of ideological work they do within a larger systems of educational practices. By identifying the strengths and weaknesses of current practices, media literacy practitioners can strategize about how to incorporate sustainability into their work.

## Who Is This Book for?

The primary audience for this book is educators working in the area of media literacy, media studies, and cultural studies. However, the language and tone is free of disciplinary jargon to appeal to those who are working in the area of sustainability education, environmental studies, and social justice. Bridging these disciplines is an emerging trend, as reflected in a number of new professional associations that seek to incorporate sustainability into communications and media studies. Examples include the Ecology, Environment, Culture Network (http://www.ntnu.edu/eecn), International Environmental Communication Association (http://theieca.org), and Institute for Sustainable Communication (www.sustainablecommunication.org). Within traditional academic associations there are now numerous subcommittees dealing with sustainability and communications, including subgroups within the National Communication Association, International Communication Association, and Western States Communication Association.

Though there is no specific book that covers both media literacy and sustainability literacy, there are several recent works that are correlating environmental issues with the media, including *Greening the Media* (Maxwell & Miller, 2012), *Climate Change and the Media* (Lewis & Boyce, 2009), *Environment, Media and Communication* (Hansen, 2009), and *Communicating Nature* (Corbett, 2006). In the realm of sustainability education, coming closest to my approach are *Critical Pedagogy, Ecoliteracy, and Planetary Crisis* (Kahn, 2010) and *The Ecology of Learning: Sustainability, Lifelong Learning, and Everyday Life* (Blewitt, 2006). The elements that make *Greening Media Education* unique are how seemingly disparate realms of education and academia are woven together, including emerging digital media pedagogies, ecoliteracy, ecologically oriented communication theory, media studies, media literacy, and sustainability education.

## Navigating the Book

In Chapter 1, I introduce a theoretical framework and overview to explain my path to gaining insight into the meaning system of media literacy. In doing so, I first provide a summary of the relationship between media literacy and sustainability, and I propose a model of green cultural citizenship. In subsequent subsections, I outline the theoretical roots of my study, starting with a discussion of how ecocriticism enables us to understand the way in which discourses, environmental ideologies, and metaphors construe the taken-for-granted world of media literacy educators. I then discuss how these discourses work within situated contexts by introducing the model of information ecologies.

Chapter 2 compares and contrasts mechanism with ecology and explores in-depth the use of metaphors in media studies and media education with a focus on the historical uses of media and environmental metaphors. Chapter 3 takes a field walk through the media ecosystem by examining various views of contemporary media, comparing and contrasting the traditional approach based on the industrial mass-media model with emerging concepts of media based on social networks and the internet. I explore how prevailing views of legacy media have dominated media studies and examine alternative perspectives that seek to reframe media disciplines from an ecocentric point of view.

In Chapter 4, I map the media literacy ecosystem's positions and discourses. I start by reviewing how media studies have impacted the major debates in media literacy. Because media literacy debates have been explored exhaustively elsewhere, my overview is selective in order to focus on those elements that relate to the problem of greening media literacy. I then summarize my research into the worldview of different media literacy organizations and practitioners, including a detailed breakdown of my data collection process and analytical methods.

In Chapter 5, I discuss the implications of my research findings, which reveal the dominant paradigm of the media literacy ecosystem. Here I describe the implicit assumptions found in media literacy documents about media, implicated actors, lifeworld, public sphere, and literacy practices. I then discuss and critique these results from an ecocritical perspective to show how mechanism dominates mainstream thinking about media literacy. I also discuss what I perceive to be ethnocentric thinking in common media literacy practices that hinder a sustainability cultural perspective. Next I discuss the barriers and opportunities for bridging media literacy and sustainability education. I close the chapter with a number of proposals for future action.

Chapter 6 clarifies my underlying assumptions about learning, cognition, communication, and sustainability education. I offer a short overview of ecoliteracy and sustainability education, and then bridge those concepts with media literacy. I then propose an alternative green media literacy framework called ecomedia literacy. I close the chapter by discussing a case study in which I implemented the ecomedia literacy framework in a classroom situation.

In Chapter 7, I conclude by highlighting trends from emerging social movements, digital storytelling, and sustainability communication that offer new pathways for media education. Here I argue that emerging media literacy practices can be leveraged as a kind of sustainability media education.

## Situating My Worldview: An Autobiography of Learning

This book is motivated by a desire to solve the enigma of why it is that media literacy educators are generally disconnected from ecological issues. Another desire is to search for common ground between media education and sustainability education, to discover a point of contact between these worldviews. I do not believe sustainability is just another pet cause for media literacy educators to take on. Like racism, gender, identity, violence, and other key social concerns of media literacy practitioners, our living systems are entitled to the attention and care they deserve. As I use the term in this book, *living systems* comprise the flow of energy and matter that sustain all life. According to Capra (2008), living systems include individual organisms, social systems, and ecosystems. I use the term instead of *environment* to avoid the dichotomization that results from the differentiation between humans and nature/environment in everyday language use. As ecocritics and ecological communication scholars have diligently noted, binaries that distinguish between *humans* and *nature* create an artificially constructed barrier that obscures how humans are embedded within and are a part of living systems (Buell, 2001; Cantrill & Oravec, 1996; Corbett, 2006; Coupe, 2000a, 2000b; Garrard, 2004; Milstein & Dickinson, 2012). A theme throughout this book is how deeply engrained compartmentalization is in our metaphors and thinking about media and living systems. But because our biosphere is endangered, we are in danger. Therefore, this is not just an intellectual curiosity, but also a desire to heal a broken paradigm that has led to the ecological disconnection I experience in my field of practice. Not only do I want to feel connected—to be part of a "tribe" of media educators who care about the environment—but I also want to be part of the solution.

As my own background and experience create the context for this book, it is useful to explain it in more depth. Clarke (2005) asserts that a situation of inquiry results in an interactive ecological system where both the researcher and research topic are embedded and situated within the unit of analysis. For example, the media literacy ecosystem is a bounded system that I have defined for the purpose of my analysis. Though based on empirical phenomena, the choice of metaphor, data sites, methods, and epistemological strategies are particular to this inquiry. These choices arise from my unique position within the media literacy ecosystem. As a media literacy practitioner trying to "green" media literacy practices, I believe that the promise of media literacy to encourage cultural citizenship has not been fulfilled because my community of practitioners is generally not addressing sustainability. Moreover, as an active participant in the media literacy ecosystem, I have had direct contact with the situation at hand and have actively written and advocated for my particular view (López, 2008; 2010; 2011a; 2011b; 2011c; 2012; 2013). It is therefore imperative that my position be explicit.

The impetus for developing an educational philosophy that combines media literacy and sustainability education can be traced to my childhood experience in alternative education. My formal education through eighth grade was at a public alternative school in Los Angeles that was founded the year I entered first grade, 1973. Area H Alternative School (AHAS) was an activist- and hippie-inspired project that became part of a network of alternative schools within the Los Angeles Unified School District. The core philosophy was distinctly andragogic: there was no boundary between students and teachers, no formal curriculum, and no requirements. You could take skateboarding class or Dutch, play baseball all day (as I did) or learn to dance the hustle. Science, math, reading, and writing were also available, but the school had no rules or requirements. Sometimes we had class in outdoor Fuller domes, and our "campus" annually migrated around the city until we acquired a permanent location five years after its founding. Students were expected to decide on their own whether or not to pursue any objective. The natural tendency was to just play. As is the case with several of my peers who went to Waldorf schools, most of us had enriching experiences but we are also somewhat "maladjusted." In other words, we have difficulty fitting into mainstream society and tend to have perspectives that are marginalized within our given professions (in my case, education).

I am certain that becoming a freethinking, independent individual is a by-product of such an environment. One of my earliest epiphanies occurred when I was ten and was changing flights in Chicago by myself (my family traveled a lot when I was younger, and sometimes I had to fly alone). I recall watching a man in the concession who could not solve a simple problem. I realized in that moment that just because someone was older, it did not mean he or she was smarter. A year later in 1977, at age 11, I transferred from AHAS to a "normal" public school where my desk, friends, and playtime were assigned to me. I did not take well to the structure the school tried to instill. After being forced to line up and recite the Pledge of Allegiance every morning for a week, I cried and feigned illness in order to go back to my beloved alternative school. But the choice to go to the mainstream public school initially had been mine, and perhaps an early glimpse of reflective thinking that was a by-product of alternative education. At the time I believed that AHAS was hindering me (my best friend was in the normal school and was accelerating at reading and writing) and I desired some systematized education. I was barely literate; my mother ended up teaching me how to read at home. Not surprisingly, at the normal school my new desk-mate had perfect handwriting and wore clean clothes. The insecurity I felt was overwhelming; fitting in within the formal education environment was difficult. Nonetheless, when I returned to the alternative school that same year, I continued to struggle. Though I had a workbook assignment due every Friday, I became a hypochondriac instead of doing the work. That year I faked illness every Friday; the following summer I worried that I would not graduate to the next grade. I had no advisors or supporters in my process; I was isolated and alone. Even though I did pass to the next grade, ultimately I continued the same behavior as before without any input from my elders and teachers. I was free, but silently stressed out.

A key aspect of AHAS was that it was a magnet school, which meant that students were bussed in from different communities to reflect LA's diversity; the fact that we were in LA meant the school was already multicultural. It was also full of colorful teachers and students, which probably was a natural result of progressive-era parents who voluntarily chose to enter their kids into this "kooky" school. In 1980, I opted for a more rigorous four-year preparatory high school, but it had experiential components as well. In addition to being an international school set in a beautiful natural environment (Sedona, Arizona), Verde Valley School emphasized an anthropological curriculum.

Each year students went on orientation camping trips and participated in two-week project periods where we could explore art, mountaineering, or study esoteric books such as *The Dancing Wu Li Masters* (Zukav, 1979) with the math teacher. Most importantly, for a three-week period during the school year we were required to live and work in an unfamiliar environment. My most formidable experiences were living with a Hopi family one year, ski backpacking in Yosemite another year, and working in a convalescent hospital in inner-city Tucson. In 1984, I traveled across country by Greyhound bus to intern at the Appalachian cultural center, Appalshop, in Kentucky during an independent senior-year trip.

More significantly, though, during my high school years in the early 1980s I participated in the most momentous intentional community of my life: punk rock. It was there that I obtained an informal education in underground politics, art, and media production. There was a do-it-yourself/do-it-with-others (DIY/DIW) ethos in our community to actively engage in political discussions and to make our own media (music, magazines, records, radio shows, books, clothes, etc.). Not surprisingly, the environment where I first encountered punks was at the alternative school. To become mediamakers in the punk movement we learned from each other, relying on informal, peer-based educational practices similar to the vision of a *deschooled society* described by Ivan Illich (1971). The experience enabled me to later become a professional mediamaker, empowering me to work in the 1990s as a journalist and multimedia producer throughout my twenties and thirties. In addition to being a mediamaker, during my punk days I was also deeply involved in community activism, energetically participating in anti-war, nuclear disarmament, environmental and solidarity movements.

My undergraduate years were characterized by more educational experimentation. In 1987, at UC Berkeley I entered into the Peace and Conflict Studies program, which was interdisciplinary by design. The department offered only a few core classes; the others were farmed out to the rest of the university based on a student-designed learning plan that culminated with a thesis project. For me, the most important class was the department's required epistemology course, which entailed reading Thomas Kuhn's (1996) *The Structure of Scientific Revolutions*. This led to deconstructing the university's role in the military-industrial complex and its ideological support for the neoliberal project that emerged under Presidents Ronald Reagan and George Bush. The interdisciplinary nature of our program enabled me to get the best the university had to offer, while at the same time helping me to gain an important counter-perspective on the hegemonic structure of university.

Punk's most important lesson for me was the significance of learner-directed DIY/DIW. Despite having no formal training in media, my experience in the punk movement enabled me to work as a media professional from 1990 through 1999 in Santa Fe, New Mexico. I worked in book publishing, film production, and magazine distribution. I also dabbled in freelance magazine writing and newspaper journalism. In addition, I was an early adapter of the world wide web and was part of a pioneering group of web developers working in northern New Mexico. The experience of on-the-job training in media has impacted my career in education, which began in 1999. Through my experience in media, I understood that I learn by doing. As an educator, this applied to the process of learning how to teach, but also as a teaching method. Correspondingly, I realized that process needs theory, which was necessary to flesh out what already had been intuitive experimentation. This correlates with the need to balance *analytic knowing* and *primary knowing*, which I understand as the harmonization of the different cognitive processes of the brain (Senge, Scharmer, Jaworski, & Flowers, 2005, p. 98). While the left hemisphere is language based and discerns rationally, the right hemisphere perceives patterns and wholeness (McGilchrist, 2009). We need both to function.[1]

When I turned 33 in 1999, my immune system collapsed, which forced me to quit my stressful job as a newspaper arts reporter and to seek out a new path. As part of this process, in 2000 I did the pilgrimage to Santiago de Compostela by walking 600 miles across northern Spain. Two months of meditative trekking allowed for events and thoughts to present themselves organically. The idea of teaching started as a daydream I was acting out during one of the long and tedious days of hiking across the plains of Castilla-Leon. For whatever reason, I started to rehearse a commencement speech for current graduates of my old high school. I imagined myself at the podium on a beautiful day in May offering tidbits of advice for young graduates as they cast out into their future from the womb of boarding school. The fantasy alternated with my real-life experience as a graduating senior in 1984 standing at that very podium reading a story from the *Tao of Pooh* (Hoff, 1983) about the emperor who kept wanting to transform into something more powerful

---

[1] I am aware that new research shows that the brain does not clearly divide tasks between each hemisphere, but the left and right brain metaphor for different styles of cognition is still relevant.

than himself (such as the sun), only to learn that there was always something else equally as strong (the clouds, wind, or rain). My contemporary self was channeling my old self, but inflecting the story with advice about exploration, seeing the world, taking time off, working new jobs, and exploring identity. The idea of encouraging formal education never entered this imaginary commencement speech. Yet, as the days wore on I thought about the important role of good teachers and how they are just as significant as informal life experience. The thought became stronger and ever present until it was the most obvious thought of all: ditch journalism to become a teacher. It was what all my grandparents did, and suddenly it was the most natural idea that had ever flowed through me. The problem was, when I returned home I did not know where or how to begin. What I did know was that I was more interested in being a mentor than a *teacher*—I wanted to be like one of the many older friends I had in high school who turned me on to strange music, obscure books, and underground art movements.

When I came home from Spain I explored teacher certification through the state of New Mexico, but learned that it involved taking a standardized exam. When I looked at sample tests I was struck by the same kind of panic that had arisen in me when I went to the normal public school: the idea of formal and "official" knowledge did not jibe with my DIY/DIW ethos. So I started with what I knew, which is making media by hand. I offered my services to a local after-school arts program, Warehouse 21 (W21) in Santa Fe, and began working as the supervisor of a teen-published zine. I insisted they write and produce everything; I brought in "old-school" examples for inspiration. In 2001, through a connection at W21 I was offered a consulting job in the Gifted and Talented Program at the Santa Fe Indian School (SFIS), a Native American boarding school in Santa Fe that serves more than 22 tribes. At first the job involved tutoring for writing, but the more time I spent at the school, the more elaborate the projects became. I shared my publishing experience in a yearbook class and also started doing video documentation for school programs. Without having very much video experience other than some classes I took at the community college, I then began teaching video in the same program. Because it was for gifted and talented students, we assessed with portfolios instead of tests. We built the curriculum around student-designed creative projects. I worked with other student-led programs within the school, but most were cut after No Child Left Behind was implemented, because teaching for tests became the priority.

During 2000–2005 I continued to teach in nontraditional educational settings. In 2002, I was trained to use a media literacy methodology developed by the New Mexico Media Literacy Project (NMMLP, now called the Media Literacy Project, www.medialiteracyproject.org). NMMLP offered professional development trainings called a Catalyst Institute, which was a four-day intensive workshop for practitioners from different sectors of society, including schoolteachers, health workers, artists, and community activists. The training led to an important revelation: I realized that the power of media could be contested through concrete educational practices. I left the workshop enraptured, convinced that media literacy is an urgently needed toolset with the power to transform society.

In 2002, I also collaborated with NMMLP to develop a Spanish-language media literacy CD-ROM for health (the first of its kind), called *Medios y remedios* (Media and Remedies). The standalone curriculum contained 60 media literacy lessons that linked media with health issues such as smoking, body image, alcohol abuse, and violence. During this time I also came to understand that because media literacy is generally not part of official education standards, many of its practitioners use media literacy in nontraditional educational environments, such as after-school programs, churches, summer camps, and activist settings. I learned that media literacy is applicable to many informal settings and is not necessarily confined to formal education venues.

During my time at the Santa Fe Indian School, I began working intensively with more than eight different tribes in northern New Mexico; I also traveled to give media literacy and video production workshops in Native American communities in Oklahoma, California, Oregon, and Washington State. The ongoing engagement with Native American communities deepened my connection to ecology. Though much of the work was centered on commercial tobacco awareness, I was also working within an alternative epistemology that incorporates sacred ecology into everyday life. While using media literacy techniques to deconstruct commercial tobacco marketing and video production to engage in regenerative community media, I began to see the interconnection between grassroots mediamaking, community health, and the environment. Nonetheless, though I felt intuitively that media literacy was connected to the projects I was working on in Native communities and applicable to ecology, I could not find materials or practitioners who were making similar connections. It was during this period when Mander (1991; 2002) told me that he opposed my efforts to bridge media literacy with ecology on the grounds that media education made media more interesting.

In 2002 at W21, we received a $40,000 grant to run a summer education program to create a peer education project for alcohol prevention through media literacy. I hired eight teens who were taught media literacy basics and then was given the space to design a street marketing project and multimedia presentation that they could deliver to younger students at public schools during the following semester. The goal was to teach the students to be peer educators because when it comes to issues like drugs, alcohol, and sex, it is better to have young people communicate the message than adults. (As will be discussed in Chapters 4 and 5, many do not view these kinds of activities as media literacy, but as "advocacy.")

For several years, I continued to lead service-learning projects using video production and media literacy as tools for getting young people to think about how media influenced their attitudes (mainly about health issues such as commercial tobacco and alcohol) and to dialogue with those attitudes through *counter media* that challenged media-generated assumptions. Based on my punk experience, I always wanted students to "get their hands dirty" and to see their own fingerprints in the media. The video projects themselves entailed collaboration, teamwork, and artistic activities through tapping into creativity and play to engender new ideas and ways of thinking about old habits.

In 2004, I embarked upon a distance-learning program to earn a master's degree in Media Studies at the New School for Social Research in New York City. During my master's studies, I discovered that the media studies tradition had very few examples of scholars connecting media with sustainability (it was not until I entered my PhD program in 2008 at Prescott College that I discovered the field of environmental communication). In 2005, after my first year in the master's program, I moved to New York City and began teaching media literacy and grassroots video production to urban youths from a variety of backgrounds, including Afro-Caribbean immigrants, African American gang members from the Bronx, and youths in foster homes. It is important to add that during this period I also became involved with a community of Buddhist practitioners called Dharma Punx. Through this experience "mindfulness" practice became an immeasurable dimension of my learning and teaching. In addition, participating in a community of Buddhist practitioners (called a *sangha*) reinforced Wheatly's (2007, p. 102) notion that, "A primary lesson of life is that nothing living lives alone. Life always and only organizes as systems of interdependence." In a sangha one learns that spiritual practice cannot happen in isolation from daily life, and that mindfulness is both an individual and social activity that necessitates a worldview that sees all relationships as interdependent. Like the Lakota Sioux phrase, "all are related" (*mitakuye oyasin*), the Buddhist perspective is very close to what I experienced when working in Native American communities.

Simultaneously, as I tried to encourage environmental activists and educators to incorporate media education into their work, I encountered a great deal of hostility. I learned from personal experience that most media literacy and environmental educators agreed with each other that there was little connection between their respective disciplines. And even though environmental educators use media to communicate with the world, there is little indication that media is seen as an integral aspect of environmental pedagogy. Consequently, as a practitioner wanting to combine these different perspectives, I realized that I would have to design a new approach that could bridge these various areas of inquiry, the features of which are explored in Chapter 6 where I describe my model of ecomedia literacy. Part of this process led me to publish a series of essays targeting my community of media literacy educators (López, 2011a; 2011b; 2011c; 2013). This community consists of diverse practitioners with varying methods but, in general, shares a common disconnection with sustainability issues. By publishing my articles and accepting presentation proposals, several editors and conference organizers have demonstrated openness to these views; nonetheless I have encountered very little dialogue about how and why media education needs to be greened. For example, in 2011, I presented on media education and sustainability at the Media Education Summit in London. Only four people attended my talk, whereas in the adjacent room a presentation about Facebook drew an audience of more than 100. The few who did attend my presentation commented that it never occurred to them that media were connected to environmental issues. I suspect this lack of awareness partially accounted for a dearth of interest in my talk. It may also be that because Facebook was a trendy topic, it attracted a larger audience. Likewise, as stated above, I had a similar experience at the Media & Learning conference in Brussels in 2012. Understanding these responses is part of the puzzle this book tries to solve.

Meanwhile, I have attempted to incorporate a green perspective into my work as a professor of media studies in an undergraduate university environment. Since 2008, I have been working with college students on a daily basis teaching media studies courses such as Media, Culture and Society; Digital Media Culture; Advanced Media Theory; Media and Gender; Media and the Environment; Intercultural Communication; and Media Ethics. As I observe with my students, it is increasingly clear that online media and personal media gadgets are a ubiquitous part of everyday life, yet they also remain unexamined in the context of global ecology. In 2011, I prototyped the ecomedia literacy curriculum in my Digital Media Culture class (discussed in greater detail in Chapter 6). This confirmed my suspicion that students initially do not make the link between media usage and sustainability, but once their attention is drawn to the subject they become aware and interested in the connection. During the course, students examined their personal media gadgets using a systems framework that connected personal media use with global ecology.

Significantly, gadget use and social networks are increasingly more relevant to student lives than what is actually learned within traditional classrooms (Gutiérrez-Martín & Tyner, 2012). The disconnection between formal education and our progressively mediated life experience has led many experts to challenge standard educational methods (RSA, 2010), media studies, and traditional media literacy practices (Bennett, Kendall, & McDougall, 2011; Kendall & McDougall, 2012). In particular, several rebel educators who also share a background in DIY/DIW punk communities have begun to subvert and question standard educational practices, calling themselves *edupunks* (Kamenetz, 2010). I feel an affinity with edupunks and draw inspiration from their experimental efforts to push the medium of education as a kind of DIY/DIW media. Much of their alternative educational practices incorporate participatory media tools, drawing inspiration from the self-organizing and deschooled character of lifelong and informal learning encouraged by everyday digital media usage. Nonetheless, though I am impressed and heartened by many of the new media education practices emerging on the internet (which are discussed in Chapter 3), I am also concerned about the disconnection between their pedagogical potential and the lack of discourse concerning their relationship with sustainability. Though there is a great deal of discussion and innovation around bringing new media practices into traditional educational settings, these proposals and efforts generally lack a green perspective.

All of these experiences have led me to conclude that media literacy and sustainability education can and should be combined into a pedagogical framework. Additionally, these experiences have also led to several biases. First, I believe in experiential education and favor informal educational environments. Second, I feel that education should revolve around communities of practice that involve peer and lifelong learning. Third, I distrust the education system and look for inspiration from outside official educational practices. Fourth, I understand that interdisciplinary education is essential. Furthermore, I also believe the role of education is related to "cultural work," which is to say that whether conscious or not, our methods and intentions propose and encourage certain ways of engaging the world. I take this position based on Walter Benjamin's (1970) concept of *author as producer*, which was an argument that cultural producers such as professional writers should be conscious of whether or not they engage their crafts as part of a larger project of criticism and political activism. In Benjamin's time, cultural work was primarily seen as a politically progressive activity as distinguished from the kinds of production for profit in which culture industries engaged. Similarly, Antonio Gramsci (Gramsci, Hoare, & Smith, 2005) developed the idea of the *organic intellectual*, which is the concept that the ruling class produces native intellectuals who reproduce the interests of their class. He theorized working-class intellectuals could advance the cause of worker rights and revolution. Combining the insights of Benjamin and Gramsci, I believe that I am a cultural worker and organic intellectual rooted in both media literacy and sustainability education.

## Epistemology

Action research pioneers Greenwood and Levin (2007) assert that in social sciences there are two general research frameworks: logical positivism and hermeneutics. Logical positivism views the world as an objectively given reality and that objective research methods can reveal truth. In contrast, I subscribe to the alternative research tradition of hermeneutics, which "is based on the ontological position that the world is only available subjectively and the epistemological project is to negotiate interpretations of this subjective world" (p. 56). In addition to hermeneutics, Heron and Reason (2006) propose that in order to encompass different ways of knowing, researchers can draw from an extended epistemology, which is knowledge that reaches beyond traditional academic theories. These extended epistemologies include experiential knowing (immediacy), presentational knowing (performance), propositional knowing (about), and practical knowing (how to) (Heron & Reason, 2006).

Given that an ecological perspective is by nature multifaceted, it is important to acknowledge that complexity is also part of my extended epistemology. Thus, my orientation is hermeneutic in combination with extended epistemologies that encompass complexity. This means "not everything in reality is socially constructed, and social constructions and conversations are not floating in an ocean of chaos...but are embedded/entangled with attractors, i.e., intervening variables at physical, chemical, biological and ecological levels of reality" (Kagan, 2011, p. 20). Subsequently, my inquiry process included meditations in nature, testing ideas in classrooms, traveling internationally to engage practitioners at conferences, walking though the streets of Rome, conversing in online social networks, gardening, worrying about the future of my children, absorbing cross-cultural encounters from past educational experiences, struggling with colleagues, and using art for inspiration.

In my effort to be true to mindful inquiry (Bentz & Shapiro, 1998), in the process of writing I intended to be honest, forthright, and authentic, attempting full disclosure whenever possible and to be mindful of what drives my own questions and definitions. Furthermore, my ecological perspective is intertwined with Buddhist practice and the experience of living and working in Native American communities. I tried to be faithful to the best of my abilities to the spirit of academic research while being true to the various strands that make up the web of perspectives that comprise this book, which ultimately is an act of media creation.

CHAPTER ONE

# Media, Environment, and Education

Whether in the form of a lecture, participatory workshop, or online course, teaching is not only a kind of communication practice but is also a kind of media that involves choices about how to frame and communicate knowledge. A university based on lecture halls structures a particular communication approach, whereas outdoor classrooms or community gardens provide alternate pedagogical environments that allow for differing forms of mediation. In a formal educational setting that has strict standards and testing requirements, the curriculum's parameters have to conform to the constraints of a particular classroom environment, including the subject matter of the course and the imposed requirements of the state. An informal setting, such as an after-school program or community arts center, affords different frameworks without the constraint of official standards.

Not surprisingly, there are similarities between education and media in how knowledge is conveyed, in particular how both have traditionally been seen as *transmissive*. The transmissive model is essentially linear: information moves from source to receiver, like a TV network broadcasting to a mass audience or an expert teacher lecturing to students. Transmissive education and mass media mirror industrial production and distribution. As linear systems, they reflect a 19th-century concept of knowledge in which information moves through Cartesian space. By contrast, media are now increasingly more networked and nonlinear, which in turn is leading to new educational practices.

Whether based on 19th-century or 21st-century practices, teaching and media are examples of *meaning design*: media and pedagogy are both efforts to create contexts that generate certain kinds of value. In 1996, an influential group of education experts, the New London Group, formulated the *design of meaning* as a way to redefine literacy practice (Cazden, Cope, Fairclough, Gee et al., 1996). They were responding to how multimedia, knowledge work, multiculturalism, and globalization have disrupted traditional literacy education, arguing that "educational research should become a design science, studying how different curricular, pedagogical, and classroom designs motivate and achieve different sorts of learning" (p. 73).

The concept of meaning design can serve as a heuristic for thinking about greening media education, because the design of systems is often what sustainability advocates focus on. It is at the level of systems design that social and cultural practices are encouraged (Ehrenfeld, 2008; Meadows, 2009; Senge, 2008). As O'Sullivan and Taylor (2004, p. 3) assert, sustainability educators "are not purveyors of knowledge. We are designers and participants in environments and processes through which people are able to learn toward an ecological perspective."

By conceptualizing literacy as meaning design, the New London Group (Cazden et al., 1996, p. 82) makes an obvious, but important, point.

> Any successful pedagogy must be based on views about how the human mind works in society and classrooms, as well as about the nature of teaching and learning. While we certainly believe that no current theory in psychology, education, or the social sciences has "the answers," and that theories stemming from these domains must always be integrated with the "practical knowledge" of master practitioners, we also believe that those proposing curricular and pedagogical reforms must clearly state their views of mind, society, and learning in virtue of which they believe such reforms would be efficacious.

In this regard, it is essential to clarify whether or not media literacy practices and sustainability education belong to different systems of meaning design, and to compare and contrast assumptions about citizenship, learning, cognition, communication, and ecology.

## Media Literacy and Sustainability

A widely used definition for media literacy was developed by the Aspen Institute in 1993, which defines it as "the ability to strategically access, analyze, evaluate, and produce communication in a variety of forms" (Aufderheide, 1993). This is the foundation for a variety of orientations that fall under two approaches: functionalist and critical (Gutiérrez-Martín & Tyner, 2012). Functionalist media literacy teaches practical skills for how to read media messages and is often linked to information literacy. It is generally apolitical and does not promote any particular kind of activism. Critical media literacy, on the other hand, acknowledges that media play a significant role in defining power relationships within society. This approach is usually associated with activism and is not neutral when analyzing media messages. Though I sympathize with critical literacy, it can often be abused for the purpose of promoting a *protectionist* agenda, which views media audiences as powerless or as victims. Protectionists try to *inoculate* students from potential harm caused by media corporations or advertisers. The primary method of most media literacy approaches is deconstruction, which involves teaching students how to analyze media messages, such as advertisements. A variety of practitioners have used media literacy to tackle social issues such as racism, gender identity, obesity, and smoking prevention.

Many media literacy practitioners have been influenced by media studies, which historically have defined the parameter of issues related to the impact of media on society. With the exception of the field of environmental communication, the ecological crisis generally has not been linked to the other social justice issues taken on by media studies, cultural studies and by extension media education. For example, in my survey of dozens of undergraduate media textbooks, media education texts, and media studies guides,[1] none of the texts had the words *ecology* or *environment* in their indices. This is not surprising given the epistemological framework of those disciplines traditionally charged with studying media; the historical divide between the biological sciences and the social sciences and humanities is well reflected in the history of media studies. As Jagtenberg and McKie (1997, p. 20) contend:

> Communication and cultural studies in their egalitarian modes parallel science's utopian and visionary aspirations.... Both participate in a common Western 20th-century intellectual journey and are still rounding similar corners: the linguistic turn where everything seemed to hang on language; the feminist sweep that transformed contents, methods, and paradigms; the self-reflexive curve where everyone had to demonstrate awareness of their own practices; and the postmodern bend where everything had to be relativized and decentered. In traveling such paths, communication and cultural studies have done more work than science, yet both need, to stay true to their respective projects' emancipatory roots, to come to terms with the environment and its ecological imperatives as the fourth dimension of social space.

This is not to pit one cause against the other, but rather to recognize, as ecofeminists have done, that issues such as social justice, racism, sexism, and environmentalism are interconnected. Nonetheless, there is a positive shift in the priorities of media scholars towards environmentally oriented media

---

[1] A sample of the key works in this survey included standard undergraduate media textbooks (Baran, 2004; Campbell, 2009; DeFleur & Dennis, 2002; Dominick, 2009); and standard media studies textbooks (Devereux, 2007; Hartley, Montgomery, Rennie, & Brennan, 2002; Rayner, Wall, & Kruger, 2004).

research and advocacy, as exemplified by the formation of the International Environmental Communication Association, Ecology Environment Culture Network, Institute for Sustainable Communication, and subgroups within the National Communication Association, International Communication Association, and Western States Communication Association. Additionally, I have observed that events such as Hurricane Sandy in 2012 along the northern United States' Atlantic seaboard, and mounting climate disruption data, are starting to shift perception regarding environmental issues among my colleagues. The question remains: Will this trend extend to the practice of media literacy education?

Though media literacy lacks a general connection with ecology, in spirit many of the goals and aspirations of media education are in alignment with the cause of sustainability. As Blewitt (2009) proposes, media literacy and environmental education have in common the goals of participation, action, and critical engagement. So why are these disciplines separated? An important barrier has to do with the perception of the purpose of media and environmental education. For example, some assume that environmental education is primarily "nature"-based and takes place outside the context of technology (Traina, 1995). But Kahn (2010, p. 6) argues traditional environmental education approaches end up lacking "rigorous training in theoretical critique and political analysis, choosing to focus instead on the promotion of outdoor educational experiences that all too often advance outdated, essentialized, and dichotomous views of nature and wilderness." As a result, environmental education has been viewed as something done away and far from civilization, such as outdoors or gardening programs. Sustainability is often pitted as the opposite of technology, so disciplines that are viewed as technologically oriented, such as those dealing with media or computers, can be regarded as inherently anti-environmental (Bowers, 2000; Glendinning, 1994; Mander, 1991; Sale, 1996). True enough, it is difficult (but not impossible) to study media without engaging technology. However, an ecological critique of technology should be a central job of media educators, and not just the territory of so-called Neo-Luddites. While experiential nature initiatives certainly remain an important aspect of sustainability education, it is also important to be ecologically literate about the primary environment that we engage with on a daily basis: media. As such, Kahn claims alternative environmental education approaches are attempting to "more robustly link forms of environmental literacy to the need for varieties of social and cultural literacy" (2010, p. 11).

Media literacy could highlight how on a daily basis we encounter the interrelationship between media and living systems. When we use any kind of media gadget, such as a "smart" phone, tablet PC, or desktop computer, the lifecycle of that machine is deeply connected to the global economy's impact on the environment. Our devices leave an ecological footprint through their manufacture and disposal, while all the data our gadgets access and store in the "cloud" also physically impact the environment. Major environmental problems with media include e-waste, contamination, loss of biodiverse habitats, damaged health, and excessive $CO_2$ emissions (Alakeson, 2003; Greenpeace International, 2010; Leonard, 2007; Lewis & Boyce, 2009; Maxwell & Miller, 2012; Tomlinson, 2010).

Then there is media's *mindprint*, which is the way that communication influences how we define and act upon living systems (Corbett, 2006). Aspects of how media shape and define our experience of the world include (a) propagating an ideology of unlimited growth, (b) reinforcing the view that nature is separate from humans, (c) marginalizing alternative ecological perspectives, and (d) favoring industry discourses surrounding environmental issues (Beder, 1998). For example, when it comes to specific beliefs about the environment, data suggests that there is an important relationship between environmental perception and media exposure. In a study of the correlation between consumption in the United States and advertising, Brulle and Young (2007) highlight that $971 per capita in ad dollars were spent in the United States in 2005, and that from the period between 1900 and 2000, there was a direct correlation between advertising dollars and increased consumption. As of 2001, according to the National Environmental Education and Training Foundation's study of environmental literacy in America, 63% of people were informed about the environment from television (Coyle, 2005). Subsequently, Hansen (2009, p. 3) attests,

> While the roles of formal education in acquainting us with the public word and image vocabulary of the environment should not be overlooked, much, maybe most, of what we learn and know about "the environment," we know from the media, broadly defined. Indeed, this applies not only to our beliefs and knowledge about those aspects of the environment, which are regarded as problems or issues of public and political concern, but extends much deeper to the ways in which we, as individuals, citizens, cultures and societies view, perceive and value nature and the natural world environment. Not only has our mainstream media model co-evolved with the system of advertising, consumption, and the ideology of unlimited growth, but the rise of global mass media clearly parallels the increasing destruction of our biosphere.

Another dimension of media's mindprint is the phenomenological experience of how media impact our sense of place, space, and time. This area of inquiry has traditionally been the focus of the field of media ecology, which views media as primarily technological environments (Lum, 2006). Its practitioners often use the term *ecology* according to a technical definition in which it represents a *system of systems* as opposed to the conventional understanding of ecology as shorthand for the interconnected system of biological communities and their relationship with the environment. Though media ecology is not explicitly correlated to living systems, many of the celebrated scholars at the core of the media ecology tradition (Ellul, 1964; McLuhan, 2002b; Mumford, 1967, 1970; Ong, 1982; Postman, 1993) were critical of modern media technology and longed for a return to less technologically complex times. They argued that technology and media alter our cognitive environments: they shape not *what* we think, but *how* we think. One example is Mumford's (1967) discussion of how the advent of mechanical clocks changed our perception of time. Another example is the idea that the alphabet and print media have reconfigured our mental patterns to favor abstractions over embodied experience (McLuhan, 2002a). Abram (1996) combined the insights of media ecology with phenomenology to assert that alphabetic literacy has impacted how Western culture experiences the natural world. More recently scholars have attempted to understand the influence of mobile gadget technology on our awareness and experience (Moores, 2012), which further complicates how we define the relationship between media and space, place and time.

As Naughton (2012, pp. 1-2) illustrates, it is useful to recall the root of the metaphor from the sciences to understand that media are more than their content, but are an environment.

> The conventional, everyday interpretation holds that a medium is a carrier of something. But in science, the word has another, more interesting, connotation. To a biologist, for example, a medium is a mixture of nutrients needed for cell growth. And that's a very interesting interpretation for our purposes. In biology, media are used to grow tissue cultures—living organisms. The most famous example, I guess, is the mould growing in Alexander Fleming's Petri dishes which eventually led to the discovery of penicillin. What I want to do is apply that perspective to human society: to treat it as an organism which depends on a media environment for the nutrients it needs to survive and develop. Any change in the environment—in the media which support social and cultural life—will have corresponding effects on the organism. Some things will wither; others may grow; new, mutant, organisms may appear. The key point of the analogy is simple: change the medium, and you change the organism.

This environmental perspective—which sees media as a technological or ideological environment—is increasingly crucial for understanding the complexity of media's role in society. Moreover, as I have argued in the past (López, 2012), the everyday consumption and use of media gadgets takes place within a planetary *media ecosystem*: the ecologically embedded sum of all our technologically mediated interactions on planet Earth. This media ecosystem includes physical, sociocultural, and cognitive ecosystems, such as the lifecycle of gadgets and energy used to power the system (physical ecosystems), civic and symbolic realms (sociocultural ecosystems), and the phenomenological experience of time, space, and place (cognitive ecosystems). The essential observation here is that *media are an environment that grows culture*.

## From Media Literacy to Ecomedia Literacy

A classroom activity found on numerous media literacy websites, "American Alphabet," involves juxtaposing two graphics in a PowerPoint presentation (see Figure 1).[2] The first slide is a group of pictures taken of plants from a local ecosystem. The second is a "ransom note" collage of cutout letters from common product logos that spell out the English alphabet. Students are asked to identify the names of plants in the first slide, and then they are asked to name the brands associated with the cutout letters in the second one. When I have done this exercise, students typically do not know the names of common plants in their local ecosystem, yet they have no problem identifying the brands in the collage. I am assuming that most teachers who do this exercise have similar results.

Depending on the context of the presentation and the facilitator's goals, the follow-up discussion will be guided by how *environment* and *media* are framed. For example, in a traditional media literacy setting, juxtaposing images from a local ecosystem with those from the socio-technologically constructed realm of media could communicate that both are a kind of environment, but that they are disconnected from each other. This disconnection could be reinforced if the point of the exercise is not to discuss ecological awareness, but simply to demonstrate the lack of it and to use the "natural world" as a kind of "negative space" to highlight what we know about media. The activity could also lead to the presumption that media are primarily a symbolic environment composed of corporate brands.

---

[2] Carrie McLaren authored the original curriculum that features this exercise (www.stayfreemagazine.org/ml/index.html).

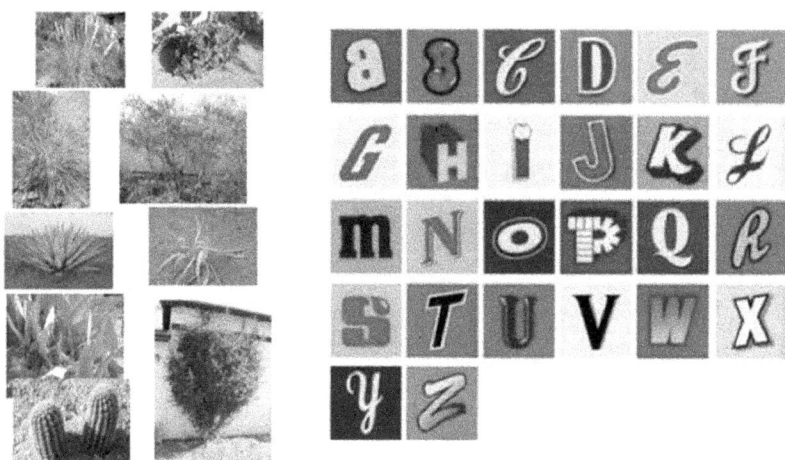

Figure 1. Images from the "American Alphabet" activity. Students are first shown a slide of the images on the left and are asked to identify plants in their local bioregion (such as those pictured here). They are then shown a slide of the image on the right and are asked to name the brands found in the cutout letters. The goal of the activity is to discover which "environment" they are more familiar with. Plant photos by Antonio López; brand alphabet collage by Heidi Cody (www.heidicody.com). Used with permission.

This discussion would be different from the perspective of ecomedia literacy, which has green cultural citizenship as its primary framework. According to *The Merriam-Webster Dictionary*, citizenship is defined as "the quality of an individual's response to membership in a community." In regards to media, cultural citizenship expands this notion to encourage active engagement with the public sphere. In contrast, economic citizenship is primarily characterized by passive consumerism and market fundamentalism. In the context of sustainability, cultural citizenship can be greened by an *eco-ethical* orientation. As Maxwell and Miller (2009, pp. 19-20) highlight:

> Economic citizenship predicated on limitless media growth diminishes potentially egalitarian and sustainable production, consumption, and participation, because it omits the impact on climate change of media technology and uptake.... Green citizenship looks centuries ahead, refusing to discount the health and value of future generations as it opposes elemental risks created by capitalist growth in the present. This necessitates an eco-ethical orientation towards the media.

By implication, green cultural citizenship calls for improving public discourse. This can be encouraged by traditional media literacy approaches that focus on analyzing media texts as a way of promoting critical literacy and a better-informed citizenry. But textual analysis also needs to be supplemented by technological literacy that accounts for the physiological impact of media gadgets on living systems and examines how media affect our sense of place, space, and time. Furthermore, an eco-ethical orientation entails *ecocentricism*, recognizing that "human beings live in a *more-than-human-world*, of which they are only one part" (Curry, 2006, p. 46, emphasis original). By contrast, economic citizenship is anthropocentric, or *human-centered*. A green cultural citizen is ecocentric by embodying sustainable behaviors and cultural practices that shape and promote ecological values. This corresponds and is enhanced by Thomashow's (1995, p. 139) holistic notion of *ecological citizenship*:

> The ecologically aware citizen takes responsibility for the place where he or she lives, understands the importance of making collective decisions regarding the commons, seeks to contribute to the common good, identifies with bioregions and ecosystems rather than obsolete nation-states or transnational corporations, considers the wider impact of his or her actions, is committed to mutual and collaborative community building, observes the flow of power in controversial issues, attends to the quality of interpersonal relationships in political discourse, and acts according to his or her convictions. The ecologically responsible citizen recognizes that he or she lives a life in nature, in conjunction with other people, in the common interest. Where does one practice this approach to life if not in the common domain?

Green cultural citizenship promotes sustainable cultural practices. Cloud (2010) asserts, "A practice (or set of practices) is unsustainable when it undermines the health of the very system upon which it depends and therefore cannot be continued over time" (p. 168). By contrast, "a sustainable practice enhances the health of the systems upon which it depends by creating favorable conditions for it to thrive indefinitely" (p. 168). It is my core belief that doing this means shifting our primary mental models from 19th-century mechanism to 21st-century ecology. According to Capra (2008, p. 366), the key characteristics of 19th-century perception include:

> the view of the universe as a mechanical system composed of elementary building blocks, the view of the human body as a machine, the view of life in a society as a competitive struggle for existence, the belief in unlimited material progress to be achieved through economic and technological growth and—last but not least—the belief that a society, in which the female is everywhere subsumed under the male, is one that follows from some basic law of nature.

By contrast, Capra believes we are in the midst of a *paradigm shift* based on ecological awareness, which "recognizes the fundamental interdependence of all phenomena and the embeddedness of individuals and societies in the cyclical processes of nature" (2008, p. 366).

Unlike the traditional media literacy approach that focuses on the study of texts, symbols, and messages as separate from living systems, in my view, in order to encourage green cultural citizenship, ecomedia literacy should support learners to:

- reconnect an awareness of media with their physiological impact on living systems;
- recognize media's phenomenological influence on the perception of time, space, place, and cognition;
- understand media's interdependence with the global economy, and how the current model of globalization impacts living systems;
- analyze how media form symbolic associations and discourses that promote environmental ideologies; and
- become conscious of how media impact our ability to engage in sustainable cultural practices by encouraging new uses of media that promote sustainability.

Ultimately, the goal of ecomedia literacy is to encourage mindfulness for how everyday media practice impacts our ability to live sustainably within earth's ecological parameters for the present and future. In doing so, it promotes the understanding that media as a whole are a socio-technological ecosystem embedded within living systems.

Environmentally harmful technology and an ideology of exploitation combine as an anthropocentric system of production and consumption. From this perspective, media are pedagogical: they teach us how to act upon and live within the world. Such a view corresponds with Orr's (1994) proposition that all education is environmental education—regardless if it is anthropocentric or ecocentric. The anthropocentric worldview permeates the taken-for-granted world where education policy in North America is formulated, as well as the background in which media literacy education is conceived. Media and education entail an implicit environmental worldview that is often not acknowledged or reconciled. Thus, green media literacy must start with the understanding that our global media ecosystem is embedded within earth's living systems. By embedding a view of media ecosystems within the life-support system of planet Earth, such a perspective calls for an implicit ethic of care, acknowledging how all media directly impact living systems on earth, human and nonhuman alike. It is from this perspective that I approach media literacy education.

## Conceptual Framework

Given that there is a "natural" connection between media and living systems, why has media literacy education generally failed to address this relationship? In the following sections I establish a theoretical framework to explain how I answer this query. I assert that media are generally conceived of as *disembedded* from ecology and that this perspective is part of an unsustainable, anthropocentric perspective based on mechanism. I start with ecocriticism, which enables us to identify and challenge how mechanism is perpetuated through discourse—the way we talk about things. Next, I claim that mechanism is maintained by language, in particular, through the use of conceptual metaphors. These metaphors form the imagined world of media literacy educators, and therefore play a significant role in defining how media are conceived and taught. By identifying and studying the social context in which these metaphors are used within an information ecology, I propose that it is possible to understand why media literacy education generally eschews an ecological perspective.

## Discourses

McLuhan proposed that in an electronically mediated world, we are like the fish that do not know the sea (2002b), an indication that media encompass us as an environment. Building on McLuhan's insight, environmental metaphors have been utilized by media critics, educators, and activists to describe the all-encompassing experience of inhabiting an electronically mediated world, including *media ecology* (Logan, 2007), *cultural environmental movement* (Gerbner, 1998), and *ecology of images* (Sontag, 2002). Silverstone (2007) calls this media environment a *mediapolis*, which is a *frameworld* with an implicit moral order based on globalization as a social, political, cultural, and technological phenomena. As such, Silverstone endorses "the idea of the media as an environment, an environment which provides at the most fundamental level the resources we all need for the conduct of everyday life. It follows that such an environment may be or may become, or may not be or become, polluted" (p. 13). According to Silverstone, screens delineate the boundary of the mediapolis, and just as a terminal is a place of entry and departure: "The screen is an interface, a frame, a window, a mask and a barrier" (p. 20). Given the state of our planetary ecosystems, I accept without reservation that the perception and presence of living systems in the mediapolis is of grave importance.

More recently there has been an evolving use of the environment metaphor to describe various aspects of emerging media systems. In the *blogosphere* it is increasingly common to use the term *ecosystem* to describe specific media environments, such as the *Facebook ecosystem* or *iPhone ecosystem*.[3] These technological *ecosystems* have data *clouds*, server *farms*, media *streams*, signal *fields*, information *flows* and network *feeds*, most of which can be found on the *web*. A part of it is even called *Amazon*. And in these ecosystems one can even find *bugs*, *viruses*, and *memes*. Yet when used in this technological context, the notion of ecosystem lacks any connection to living systems impacted by media, and hence represents an incomplete use of the ecosystem concept. For instance, Naughton (2006, 2012) draws heavily on the ecosystem metaphor to describe internet-based media without making any reference to living systems. The lack of awareness of the connection between media and living systems is not uncommon, in particular in the field of media studies. Stated starkly:

> The prevailing myth is that the printing press, telegraph, phonograph, photograph, cinema, telephone, wireless radio, television, and internet changed the world *without* changing the Earth. In reality, each technology has emerged by despoiling ecosystems and exposing workers to harmful environments, a truth obscured by both the symbolic power and the power of moguls to set the terms by which such technologies are designed and deployed. Those who benefit from the ideas of growth, progress, and convergence, who profit from high-tech innovation, monopoly, and state collusion—the military-industrial-entertainment-academic complex and multinational commanders of labor—have for too long ripped off the Earth and workers. (Maxwell & Miller, 2012, p. 9)

Given that the coinage of the term *ecosystem* in the 1930s is rooted in a view of living systems as interdependent communities (Golley, 1998), I believe any use of the term ecosystem without reference to living systems is deficient and problematic. As such, I prefer Lappé's (2011, p. 15) grounded definition of ecology, which is "relationships among organisms and their environment." Lest we forget, humans are animals that inhabit an environment that also

---

[3] For example, on March 8, 2013, a Google search for the phrase *Facebook ecosystem* generated 36,000 results. *iPhone ecosystem* generated 11,900 results.

includes technological systems and their modes of production, both cultural and material.

Indeed, it is important to recognize that *ecology* is not neutral, nor does it have universal meaning. Among those who call themselves ecologists there is a broad philosophical spectrum that varies from a phenomenological connection with nature as sacred (Berkes, 1999; Harding, 2006) to a scientific view that ecology is a system of systems (Allen, Tainter, & Hoekstra, 2003; Odum & Barrett, 2005). Hornborg (2001) asserts that these approaches represent the difference between *embedded* and *disembedded* worldviews, embedded being ritualistic/spiritual and disembedded representing scientific/technological approaches. According to Berkes (1999), embeddedness is the defining characteristic of a sacred relationship with nature and characterizes the worldview of many traditional and land-based peoples.

The difference in perception of ecosystems as technological or as embedded is tied to discourses about the environment. A discourse is "a shared way of apprehending the world" (Dryzek, 2005, p. 9) that expresses taken-for-granted assumptions about how the world works. Our view of the environment is closely related to how we talk about it. As Corbett (2006, p. 6, emphasis original) notes:

> "nature," and in a different way "environment," are complicated cultural concepts, not just words. Nevertheless, they *communicate*. The words and how we use them interpret and define what exists beyond humans. This is nature, that is not. This is an environmental issue, but this is not. The definitions and meanings to a certain extent influence our behaviors and practices and our communication about it.... That is not to say, however, that nature or environment or whatever we want to call it is one big social construction and doesn't really exist out there independent of us and our definition of it. The physical, nonhuman world does exist; ecosystems and their inhabitants would unfold and continue just fine without humans. *Social construction*—the definitions and meanings we come to accept through our social interaction—is just one component. Other components are the historical and cultural contexts in which we live and the unique sets of individual experience we carry with us.

Corbett's assertions are related to social constructionism and symbolic interactionism, which are tied to three assumptions about how language is socially constructed through negotiated and situated contexts. First, "social life consists of a process of communication and interpretation regarding the definition of a situation"; our way of knowing about the world and ourselves is grounded within the *symbolic order* we are born into (Altheide, 1996, p. 8). Next, we are situated within a social world that is experienced reflexively. Finally, the notion of process is key because everything is, so to speak, under construction, even our most firmly held beliefs, values, and personal commitments. What we consciously believe and do is tied to many aspects of "reality maintenance," of which we are less aware, that we have made part of our routine "stock of knowledge." (Altheide, 1996, p. 8)

## Environmental Ideologies and Mechanism

Even in the absence of an explicit environmental stance, a discourse can communicate an environmental ideology, which is "a way of thinking about the natural world that a person uses to justify actions towards it" (Corbett, 2006, p. 26). Environmental ideologies span a spectrum ranging from anthropocentric to ecocentric. According to Corbett (2006), the most extreme anthropocentric ideology is unrestrained instrumentalism, which views the natural world as a resource that should be exploited for human use. Within the domain of anthropocentric ideologies she also includes conservationism and preservationism since they ultimately favor environmental policies meant to benefit humans. On the other end of the spectrum is ecocentricism, which Corbett aligns with *ethics and value-driven ideologies* and *transformative ideologies*. Ethics and value-driven ideologies "grant nonhuman entities 'value' that goes beyond utilitarian, scientific, aesthetic or religious worth to possessing intrinsic value or inherent worth" (2006, p. 37). Transformative ideologies are associated with deep ecology, social ecology, ecofeminism, indigenous ideologies, and "Eastern traditions."

I personally align with transformative environmental ideologies that are grounded in Shiva's (2005) concept of an Earth Democracy, which is "based on the intrinsic worth of all species, all peoples, all cultures; a just and equal sharing of the earth's vital resources; and sharing the decisions about the use of the earth's resources" (p. 6). Shiva's perspective is a key element of *ecojustice*, a set of principles that include:

- The need to eliminate the toxic contamination of individuals, plants, and animals, which is also the basis of the eco-racism that occurs when the toxic wastes of industries contaminate the air, water, and soil of nearby economically poor and culturally marginalized neighborhoods, and when shipping toxic wastes across national boundaries.
- The need to eliminate the colonization of the South by the North that results, in part, from the exploitation of the South's natural resources, and from the efforts to replace their traditions of intergenerational and community self-sufficiency with the West's emphasis on an individual/consumerdependent [sic] lifestyle.

- The need to revitalize the cultural commons that represent the intergenerational knowledge and skills that are less dependent upon monetized activities and relationships—and that have a smaller carbon and toxic footprint. Also the need to conserve what remains of the environmental commons.
- The need to pursue lifestyles that ensure that future generations will inhabit viable environments that allow them to live morally coherent and symbolically rich lives. (Bowers, 2012, p. 224)

When media ecosystems are discussed solely within an economic or technological framework, I consider it to be an example of an anthropocentric discourse because it implicitly endorses the view that technology, progress, and economics are outside the domain of living systems. This includes most references to *media ecosystem* on the web. I believe an ecocentric discourse is when media ecosystems are contextualized by their impact on living systems and ecojustice. Though they are not using the term *media ecosystem* or *ecojustice* in their work, examples of ecocentric discourses can be found in media like the *Story of Stuff* (Story of Stuff Project, 2009), *Food, Inc.* (Kenner & Pearlstein, 2008), and *Avatar* (Cameron & Landau, 2009).

Media conceived of as disconnected from living systems is rooted in mechanism and tied to the dominant discourse of *industrialism*, which is "characterized in terms of its overarching commitment to growth in the quantity of goods and services produced and to the material wellbeing that growth brings" (Dryzek, 2005, p. 13). In an industrial discourse, "natural resources, ecosystems, and indeed nature itself, do not exist" (Dryzek, 2005, p. 57). Garrard (2004, pp. 16-17) relates this framework to the *cornucopia* discourse: "The key positive claim put forward by cornucopians is that human welfare, as measured by statistics such as life expectancy or local pollution, has demonstrably increased along with the population, economic growth and technological progress." Industrialism is ultimately an expression of a *paradigm*. According to Meadows (1991, p. 3),

> A paradigm is not only an assumption about how things are; it is also a commitment to their being that way. There is an emotional investment in a paradigm because it defines one's world and oneself. A paradigm shapes language, thought, and perceptions—and systems. In social interactions, slogans, common sayings, the reigning paradigm of the society is repeated and reinforced over and over, many times a day. Whenever a speaker of an Indo-European language says a sentence, nouns and verbs reinforce the paradigmatic distinction between things and processes (in some other languages there are only processes). Every time you buy or sell something, you affirm a shared paradigm about the value of money. Every time the president rejoices when the gross national product (GNP) goes up, he strengthens the paradigm of

economic growth as an unquestioned good. In general, media and education are situated within the dominant paradigm of mechanism.

## Ecocriticism

It is my contention that to examine and critically engage language is to go to the root of how we perceive the world. In this respect, a tool for challenging mechanism is to engage in a kind of ecocriticism, which is "the ability to critique existing discourses, cultural artifacts, forms and genres, and explore alternatives" (Garrard, 2009, p. 19).

> [Ecocritics] who analyze literary and other texts from an environmentalist standpoint, observe our environmental crisis poses not only technical, scientific and political questions, but also *cultural* ones. Our habits of representation affect and reciprocally reflect our actions, but the enormous temporal and spatial scale of phenomena such as climate change and mass extinction, and the complex moral questions inherent in them, pose challenges for our existing artistic forms. (Garrard, 2009, p. 19, emphasis original)

The key concern of the ecocritical approach is to recognize how the values promoted by mechanism and the cornucopian discourses are undermined by their lack of recognition of ecological limits.

## Metaphors

According to Machin and Mayr (2012, p. 221) metaphor is "the means by which we understand one concept in terms of another, through a process of which involves a transference of 'mapping' between two concepts." Bowers (2012, p. 164) describes *root* metaphors as "deep, generally taken for granted interpretative frameworks that influence thinking, values, and practices over a wide range of cultural activities—and over generations and even thousands of years." For example, the metaphor *framework* embodies a host of cultural assumptions. A frame implies a window that opens up to a kind of Renaissance-inspired linear perspective space in which we look out into the world from the vantage of a screen. Romanyshyn's (1989, p. 42) phenomenological study of linear perspective yields this important observation: "The condition of the window implies a boundary between the perceiver and the perceived.... Ensconced behind the window the self becomes an observing *subject*, a *spectator*, as against a world which becomes a *spectacle*, an *object* of vision." From an ecological perspective, this kind of orientation is problematic. As Milstein and Dickinsen (2012, pp. 513-514) note,

> Humans use the body to relate within/to nature, a practice that is culturally constructed and mediated in gendered ways. Gaze and ocularcentrism (favoring of vision) play a central role in frontally orienting humans to nature, and a number of scholars critically explore and critique frontal orientations. Foucault's Panopticon illustrates how a human one-way subjective gaze in ocularcentric cultures enables prototypically androcentric orientations of hierarchy and domination. In humanature relations, the favoring of the gaze can privilege a frontal orientation to nature that distances and objectifies.

The frame metaphor, if not properly contextualized, retains this sensibility of separation, reinforcing a Cartesian sense of space.

Root metaphors reinforce paradigms of thought, such as mechanism at the root of industrialism. Emerging during the Industrial and Scientific Revolutions and influenced by scientific pioneers like Copernicus and Newton, mechanism signifies that humans are disconnected from living systems because nature can be isolated and analyzed by examining its disconnected parts (Capra, 1983). Mechanism correlates with the root metaphor of *individualism* because the concept of an *autonomous self* means that people have internal worlds that are isolated from their environments. According to Bowers (2012), analogs that reinforce the root metaphors of mechanism and individualism include *progress*, *development*, *technology*, and *freedom*.

Both *media* and *environment* are examples of metaphors that will take on different meanings according to their associated analogs. In the case of my interaction with Mander, when I used the word *media* he connected it with an assortment of taken-for-granted assumptions about the nature and purpose of media. It is likely that he linked my use of the term with his own analogs of *technology, progress, consumer culture, loss of the sacred*, and other metaphors used in his work (1991, 2002). Likewise, when I discussed ecology and media literacy with the president of the media literacy organization in Brussels, she associated it with a set of assumptions based on different analogs for *environment*, which categorized it as not related to *media*. In both encounters we had different analogs for the metaphors of media and environment, and therefore we were communicating with different conceptual frameworks.

Bateson (2000) argued that metaphors serve as reality maps, but we often mistake the map for the territory. According to Bowers (2008, 2009, 2012), this means that we use metaphors (maps) based on previous ways of thinking that are disassociated from the current context of everyday life (territory). This results in preceding worldviews colonizing the present through language. For example, the metaphor "progress" is rooted in the 19th century when it was believed that the natural world was a resource that could be technologically exploited without limit. Sustainability advocates and ecocritics challenge the meaning of progress when it is associated with economic development and environmentally destructive technology, arguing that progress is responsible for "overshooting" the carrying capacity of living systems (Meadows, Randers, & Meadows, 2004). Reconciling progress with current environmental conditions can lead to what Bateson (2000) referred to as a double bind. A double bind results when one tries to solve a problem with the same kind of thinking that created the problem. It is likely that Mander's response to my media literacy proposal was his perception that I was advocating a double bind: engaging media to solve a problem that is partially the result of media. Subsequently, it is essential for media literacy educators to consciously engage metaphors that dominate educational practices, not just metaphors tied to the notion of progress, but those that are fundamentally tied to media and environment. As Garrard states:

> The study of rhetoric supplies us with a model of a cultural reading practice tied to moral and political concerns, and one which is alert to both the real and literal and the figural or constructed interpretations of 'nature' and 'the environment.' Breaking these monolithic concepts down into key structuring metaphors, or tropes, enables attention to be paid to the thematic, historical, geographical particularities of environmental discourse, and reveals that any environmental trope is susceptible to appropriation and deployment in the service of a variety of potentially conflicting interests. (Garrard, 2004, p. 14)

This is not to criticize the use of these metaphors or to make judgments about the kinds of metaphors we use to explain the world, but to raise awareness. Indeed, throughout this book I use many of the same metaphors deployed by media literacy educators. According to Lakoff and Johnson (1980), the reason metaphors are used in language is because they are grounded in experience. We use many *container* metaphors (such as knowledge *construction*) because we have the embodied experience of entering and leaving spaces. Subsequently, in Western culture we divide the world between interior and exterior experiences. So when we describe an abstraction (such as media), metaphors help us concretize and make sense of what we are talking about. Ultimately, the reason to study metaphors is because inevitably they include some experiences of the world while excluding others. Metaphors can illuminate and hide simultaneously.

## Figured Worlds

Media literacy education's discursive environment comprises root metaphors that shape a *figured world*. Figured worlds "are 'theories' or models or pictures that people hold about how things work in the world when they are 'typical' or 'normal'" (Gee, 2011a, p. 173). A figured world, which is similar to *folk theories, frames, scenarios, scripts, mental models,* and *discourse models*, is something with which all of us operate. We have many figured worlds depending on the context and situation with which we are involved: We have mental models for personal relationships, political affiliations, workplace practices, and so forth. An example of a root metaphor that shapes the figured world of media literacy educators is *media*, which conjures a host of assumptions and beliefs about communication, democracy, and public sphere that include and exclude certain actors within its system. These taken-for-granted assumptions work together with other root metaphors to shape a paradigm of media literacy practice. Indeed, "the institutional practices of teaching about popular culture must be understood as a technology for the naturalization of specific reading and writing practices, particular ways of making meaning and understanding the world which are far from neutral" (Bennett, Kendall, & McDougall, 2011, p. 4). As such, metaphor usage does not exist in isolation from social practice; metaphors have situated meanings in social contexts where certain values and perceptions are shared. A deeper understanding of the dominant metaphors within a social context helps clarify the kind of world that is envisioned by media literacy educators and whether or not it is compatible with green cultural citizenship.

In addition, an important characteristic of media literacy education's figured world is how *implicated actors* are discursively represented. Implicated actors are "actors silenced or only discursively present—constructed by others for their own purposes" (Clarke, 2005, p. 46). Clarke describes two kinds of implicated actors:

> First, are those implicated actors who are physically present but are generally silenced/ignored/invisible by those in power in the social world or arena. Second are those implicated actors not physically present in a given social world but solely discursively constructed; they are conceived, represented, and perhaps targeted by the work of those others; hence they are discursively present. (2005, p. 46)

In the media literacy ecosystem, a number of actors are discursively present, such as educators and students, while others are mostly implied, such as mediamakers or policymakers.

## Information Ecologies

It is my belief that integrating media literacy and green cultural citizenship requires critiquing the existing metaphor usage of media literacy educators from an ecological perspective, but also repurposing ecological metaphors to shift how we think about the field. This is because, "Taken for granted patterns of thinking and communicating are difficult to recognize and...are especially difficult to change.... The use of new analogies, in turn, leads more people to become aware of what they previously took for granted" (Bowers, 2012, p. 10). To this end, my strategy is to define the activities and practices of contemporary media literacy educators as taking place within a *media literacy ecosystem*. The concept of a media literacy ecosystem is based on Nardi and O'Day's (2000, p. 49) model of an information ecology, which is "a system of people, practices, values, and technologies in a particular local environment." In addition, my use of information ecology incorporates theories of communication ecologies (Altheide, 1995; Bateson, 2000; Luhmann, 1989), sociocultural ecosystems (Clarke, 2005; Liska & Cronkhite, 1995), and systems thinking (Capra, 1996, 2008; Meadows, 2009; Morris & Martin, 2009; O'Connor & McDermott, 1997; Pittman, 2004).

Nardi and O'Day's (2000) use of the ecology metaphor is an effort to reframe how we normally think about technology, because "our concepts of technology are often embodied in highly packed metaphors" and "metaphors channel and limit our thinking" (p. 25). As such,

> There is an urgency in the notion of ecology, because we are all aware of the possibility of ecological failure due to environmental destruction.... We feel a sense of urgency about the need to take control of our information ecologies, to inject our own values and needs into them so that we are not overwhelmed by some of our technological tools. (Nardi & O'Day, 2000, p. 56)

Depending on the technology metaphor used—*tool*, *text*, *system*, or *ecology*—each will raise different questions about who really is in control of technological development. Nardi & O'Day state:

> Metaphors matter. People who see technology as a tool see themselves controlling it. People who see technology as a system see themselves caught up inside it. We see technology as part of an ecology, surrounded by a dense network of relations in local environments. Each of these metaphors is "right," in some sense; each captures some important characteristics of technology in society. Each suggests different possibilities for action and change. (2000, p. 27)

According to Nardi and O'Day's model, examples of information ecologies include libraries, hospitals, self-service copy centers, computerized classrooms, and virtual networks. Media literacy practitioners comprise a kind of information ecology in several ways, but not exactly in the model offered by Nardi and O'Day. In their framework,

> An information ecology is a complex *system* of parts and relationships. It exhibits *diversity* and experiences continual evolution. Different parts of an ecology *coevolve*, changing together according to the relationships in the system. Several *keystone species* necessary to the survival of the ecology are present. Information ecologies have a sense of *locality*. (2000, p. 51, emphasis original)

Media literacy education is indeed a system of complex parts and relationships; the field is fairly diverse with many complementary and competing practices that evolve over time, while coevolving with particular aspects of the world (such as emerging technologies, education policy, and social concerns); media literacy education does have a keystone species that develops important documents (such as curricula and teacher resources), runs media literacy organizations, trains other practitioners, and advocates particular visions of media literacy education through books and academic articles; and media literacy education can have a sense of locality (but not always).

Nardi and O'Day's model of the keystone species of an information ecology (like librarians) has an analog in media literacy in the form of leading developers and trainers of media literacy. For example, Tessa Jolls is the president and CEO of an influential media literacy organization, Center for Media Literacy, which evolved from the publication *Media & Values*. She is a working member of the National Association for Media Literacy Education, which sponsors *The Journal of Media Literacy Education*. She has authored numerous articles, develops curricula, and trains and consults a variety of national and international organizations. Another example is Sut Jhally, the founder and executive director of the Media Education Foundation. As an author and scholar who actively participates in media literacy debates (Lewis & Jhally, 1998), he produces popular media education documentaries and was an early supporter of the Action Coalition for Media Education (ACME).[4]

My model of the media literacy ecosystem differs from Nardi and O'Day's framework on two points. First, the notion of locality in media literacy is altered by the influence of the national movement. Second, media as a technological system replaces local technology as an organizing element. On

---

[4] Jolls and Jhally did not participate in this study.

the one hand, media literacy can be very local as it is practiced within particular settings. This was very true of my personal experience and how working and teaching in different environments (such as New Mexico, New York, and Italy) impacted the way I taught and related to the media literacy movement. In addition, there are many regional media literacy organizations that focus on the needs of their communities. However, there are also discourses about media literacy education happening on national and international scales that are "placeless" in the same way that national education standards lack local contexts. While I believe regional- or community-based media literacy organizations are kinds of information ecologies, they are also embedded in larger information ecologies of national and international media literacy movements.

Information ecologies have technology as their binding element (such as machines at the disposal of librarians, doctors, or classroom teachers). Though technology is certainly a big part of media literacy practice, I assert that for media literacy educators, the metaphor of media serves as a kind of *boundary object* that ties practitioners together. Boundary objects have been theorized as objects with commonly agreed-upon symbolic properties that border professional discourses (Gieryn, 1983; Star & Griesemer, 1989). These symbolic artifacts facilitate communication between different groups, but their meanings and interpretation vary according to context. Wenger (1998) uses the example of an insurance claim form, which on the surface is pretty straightforward in terms of the kind of information it contains. He notes that the claimant and processor will view it differently, as will the floor managers who oversee claims adjusters, the corporate executives who set policy, and the mail delivery person. Yet, to all of them, the object exists as a particular form that is indisputable. *Media* are not an *object* per se, but the media metaphor is a discursive object that binds media literacy educators together.

As a boundary object, the media metaphor is *of* and *for* something, which correlates with Carey's (2009) discussion of media artifacts. He gives the example of an architectural blueprint, which is a drawing *of* a house—a literal map of the house. But it is also *for* building the house. It has an intention to do something. Likewise, media metaphors are maps *of* and *for* culture. Carey contextualizes this within communication theory:

> Models of communication are, then, not merely representations of communication but representations for communication: templates that guide, unavailing or not, concrete processes of human interaction, mass and interpersonal. Therefore, to study communication involves examining the construction, apprehension, and use of models of communication themselves—their construction in common sense, art, and science, their historically specific creation and use: in encounters between parent and

child, advertisers and consumer, welfare worker and supplicant, teacher and student. Behind and within these encounters lie models of human contact and interaction. (2009, p. 25)

The media-metaphor-as-boundary-object offers us a map of the implicit figured world that media literacy practitioners share. How media are perceived is also a recipe for action about how to teach media. As we will see in the following chapters, for the media literacy education field, the media metaphor serves as an edge and a connection between disparate practitioners.

CHAPTER TWO

# Metaphors as Meaning Design

We tend to grasp emergent realities with previously codified ways of seeing the world, which are expressed in particular metaphors we inherit from the past. McLuhan (2002b) observed this with the analogy that people view emerging media by looking through the rearview mirror. This helps explain why the initial rise of mass media was communicated with agricultural metaphors (terms such as *broadcast*, *field*, and *culture* all derived from agrarian practices), and why new digital media are often discussed from a book literacy framework (i.e., commonly used media education terms include *reading*, *literacy*, *grammar*, and *text*). Likewise, the source of the modern ecology metaphor and its influence on contemporary studies comes from the Western *silo*-ing of knowledge that splits economy and environment from their original linkage. In the 19th century when our hierarchies of knowledge were codified, the field of ecology was assigned to the biological sciences; yet the Greek word, *oikos*, at the root of ecology and economics refers to *household*, *house*, and *family*. Ernst Haeckel, who coined the term in 1866, chose *oikos* because he intended for ecology to mean the *economy of nature*. Subsequently, Jacobs (J. Jacobs, 2001, p. 10) contends that ecology's root meaning is *house knowledge* and economy represents *house management*. This integrative approach differs from the way in which economy and ecology are treated as separate realms of knowledge.

This is not to say ecological metaphors have not been used to describe human societies. On the heels of Darwin's theory of evolution, one of the founders of sociology, Herbert Spencer, applied an environmental framework on society by coining *survival of the fittest*, which inadvertently was used to justify colonialism. Human ecology emerged in the Chicago school of sociology, which studied culture and society by using ecological frameworks to investigate the relationship between people and urban environments. The Chicago school produced several scholars who extended environmental metaphors to explain the role of media and society, in particular Robert E. Park and Harold D. Lasswell. However, their approaches reflected a common theme discussed throughout this book, which is that the prevalence of environmental metaphors does not necessarily echo ecological consciousness.

For example, humans may depend on urban environments, but those environments do not end with social structures, economics, or political realities; they cannot exist without their interconnection with biological living systems. Unfortunately, it was the so-called hard sciences that were charged with that domain of knowledge. Thus, this initial split between a scientific and cultural use of *ecology* is crucial and has resulted in what Gregory Bateson (2000) calls an "ecology of bad ideas" (see below).

In order to understand how ecology metaphors have been adapted, in the following sections I explore the difference between mechanism and ecological intelligence. I then examine the importance of metaphors in terms of shaping how media are perceived by researchers and media literacy educators. I close the chapter with some suggestions for how to repurpose ecological metaphors to promote green cultural citizenship.

## Mechanism and Ecological Intelligence

The separation of ecology from the cultural disciplines coincided with the emergence of mechanism as a dominant model of the universe. Mechanism, simply put, is a reductionist model of nature that correlates with the manufacturing process (Leiss, 1972), and views the universe as composed of atomized parts that work together as one great machine. Mechanism emerged during the Industrial and Scientific Revolutions and is characterized by a Cartesian subjectivity that promotes the domination of the natural world through objectification and the positivist scientific method. According to Merchant (1989, pp. 192–193), the change from an organic to mechanistic paradigm in European culture represented a major shift in worldview:

> In the organic world, order meant the function of each part within the larger whole as determined by its nature, while power was diffused from the top downward through the social or cosmic hierarchies. In the mechanical world, order was redefined to mean the predictable behavior of each part within a rationally determined system of laws, while power derived from active and immediate intervention in a secularized world. Order and power together constituted control. Rational control over nature, society, and the self was achieved by redefining reality itself through the new machine metaphor.

Consequently, industrial-scientific societies have structured themselves to emulate machines, which Ellul (1964) calls *technique* and Mumford (1967) calls the *megamachine*. As an ordering principle, mechanism permeates all aspects of society, from economics to media systems to education, and many consider it the root of our ecological crisis (G. Bateson, 2000; Berry, 2005; Capra, 2004; Merchant, 1989).

Mechanism influences models of cognition, regarding the mind as a repository of symbolic representations based on a machine metaphor: representations move through space from person to person, and as a result individuals construct an autonomous identity that is independent from living systems. This is reflected in the prevalence of graphics that represent the mind as a set of gears, of the term "programming" used to describe how thoughts can be controlled or influenced. This machine model of the mind leads to Gregory Bateson's (2000, p. 337) concept of an "ecology of bad ideas," in which technological-scientific progress is viewed as part of a linear path of history:

> If we continue to operate in terms of a Cartesian dualism of mind versus matter, we shall probably also continue to see the world in terms of God versus man; elite versus people; chosen race versus others; and man versus environment. It is doubtful whether a species having both advanced technology and this strange way of looking at its world can endure.

The prevalence of mechanism in media literacy education relates to Giddens' (1984) idea of the *duality of structure*: the concept that institutions and the people that comprise them recursively reinforce each other. Jensen (2002a, p. 1) notes that "social subjects and social systems must be seen as continually reproducing and, to a degree, reforming each other, and they interact, not as abstract principles but in concrete practices and contexts...." For media educators, this leads to a *double hermeneutic*, the notion that we simultaneously create what we study.

In the context of Gregory Bateson's (2000) discussion of an ecology of bad ideas, media educators working within the paradigm of Cartesian dualism develop a *symmetrical relationship* with the environment. In other words, Cartesian subjectivity is a reality bubble that excludes ways of knowing that contradict its self-defined reality. This is the so-called snowball effect described by systems theory when reinforcing information within a given circuit cycles into infinite regression because it lacks *balancing feedback* from outside the system. An example of the snowball effect is illustrated by the story of Beijing's pollution problem and how people use electrically powered purifiers to clean the air in their apartments. But the coal used to power them is what is causing air pollution in the first place, so it is a solution that doesn't solve the problem, but exacerbates it. In terms of media literacy, using mechanistic models of cognition and communication will reinforce the paradigm of industrialism, remaining stuck in a system of "bad ideas"; the essential bad idea being the assumption that communication is a matter of autonomous beings transporting ideas between each other as *messages*, and that such communication is disembodied from the *thinking system* that comprises our cultural patterns and embeddedness within living systems.

An alternative framework for cognition is *ecological intelligence*. Ecological intelligence is based on Gregory Bateson's (2000, pp. 309–337) understanding that a person as not simply an autonomous *self* but is part of an interconnected *thinking system* that not only includes socially constructed knowledge but knowledge that is co-produced with the living environment. For example, we constantly respond to feedback from our actions within the surrounding world, and we adjust according to what our environment affords, thereby creating a circuit of information. Bateson's example is the lumberjack who hits a tree with an ax. Every time he strikes a blow, the information between him and the tree changes and he must adjust. The difference between each ax blow is a "difference that makes a difference," which posits information as a continuously negotiated phenomenon in which it acts more like a disturbance that perturbs a system than a static thing that moves through space.

The cognitive scientist Francesco Varela (1999) describes it differently by using the analogy of old shoes. The foot and the shoe become accustomed to each other, each adjusting for the proper fit. Likewise, our minds make constant adjustments according to environmental conditions; thus, it would be impossible to conceive of a mind independent of sensory experience shaped by its surroundings. Maturana and Varela (1998) refer to this as *structural coupling*, an example being how in the United States automobiles and cities have co-evolved to shape each other: the car and the city each have a particular structural dynamic that affects each other's development. Likewise, the human mind and its environment co-evolved and continue to do so according to prevailing conditions. To put this in a media context, consider how the content of a telephone call is insignificant as compared to how telephones (landlines and now mobile) restructure society, and visa versa. We can take McLuhan's (2002b) view that media are extensions of our nervous systems and that the kind of media that extend our cognition will shape cognition differently. As such, the nervous system that privileges technological environments will experience a deficit in feedback and experience that would be afforded by an outdoor setting. From this framework, communication is conceived differently than as a mere transmission, but as a complex experience that is conditioned by specific environments and situated practices.

Ecological intelligence has a moral dimension as well. It is one thing to see how systems interact and relate to each other; it is another to care about how and why those relations exist. As such, ecological intelligence, "is essentially relational or connective thinking, but also more than that: it is ethical, valuative, and expresses our humanity" (Sterling, 2009, p. 78). Indeed, ecological intelligence encompasses many "intelligences" that are not just intellectual. Lappé's (2011) model of the "EcoMind" includes the capacity for cooperation, empathy, fairness, efficacy, meaning, imagination, creativity, and plasticity.

Ecological intelligence is akin to systems thinking, because it is based on seeing patterns and relationships from small to large scales, and understanding interrelations between them. Warshall (2012) identifies seven schools of systems thinking, ranging from Native American systems of moral practice to systems dynamics proposed by modern scientists. What they have in common is an interrelationship between three Cs: components, connections, and configurations. As noted by Meadows (2009), the purpose of systems theory is to see the relationship between structure and behavior. In particular, systems thinking informs an understanding of how mental models structure cultural practices and the creation of institutions that impact the environment. Drawing out these relationships means seeing how the components and configurations of a system connect: "System dynamics makes clear the overarching power of deep, socially shared ideas about the nature of the world. Out of those ideas arise our systems—government systems, economic systems, technical systems, family systems, environmental systems" (Meadows, 1991, p. 2).

A visual model for systems thinking is the iceberg diagram (see Figure 2). According to this model, a main cause of unsustainable behavior is that we act upon the world at the level of events and patterns, but mental models that produce social structures influence our behaviors. Correspondingly, the mechanistic model of the world has led to the current status quo because it has generated a global economy in which its actions are disconnected from the long-term impact on living systems. Sustainability educators believe that in order to solve the global ecological crisis, we need to transform our mental models from mechanism to something related to systems thinking and ecological intelligence (Berry, 2005; Briggs, 2005; Cloud, 2010; Cloud Institute for Sustainable Education, 2011; Goleman, Bennett, & Barlow, 2012; Sterling, 2004; Stibbe, 2009; N. J. Todd & Todd, 1994).

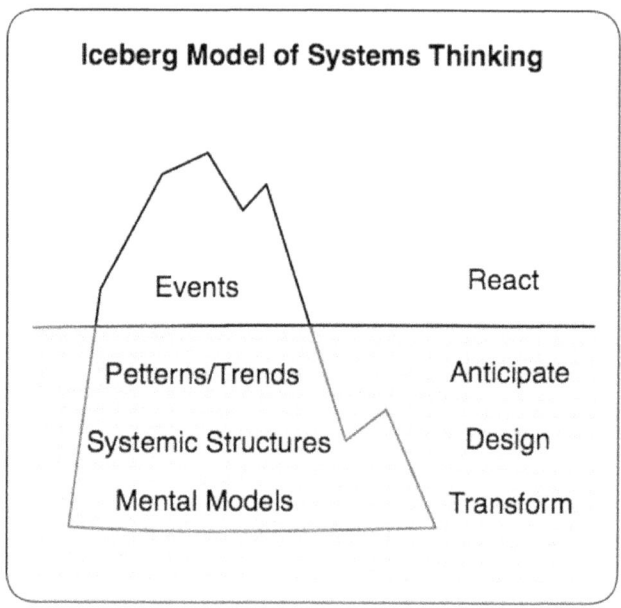

Figure 2. Iceberg diagram. This illustration shows the different levels of thought that guide our actions in the world. The aim of systems thinking is to understand how our mental models influence other thought patterns.

The distinction between mechanism and systems thinking can also be tied into Nisbett's (2004) discussion of the difference between Asian and Western mental models. Nisbett conducted psychological experiments in which North American and Japanese students looked at an animation of an underwater scene, but each saw something different. North American students focused on objects such as fish and Japanese students paid attention to the background environment. Nisbett asserts that this experiment demonstrates a primary difference between how Westerners and Asians schematize the world (see Figure 3). Westerners see the world embedded in their cognitive processes, whereas for Asians cognitive processes are embedded in the environment.

For Nisbett, this leads to the following generalizations about how the two are different:

- Patterns of attention and perception, with Easterners attending more to environments and Westerners attending more to objects, and Easterners being more likely to detect relationships among events than Westerners.
- Basic assumptions about the composition of the world, with Easterners seeing substances where Westerners see objects.

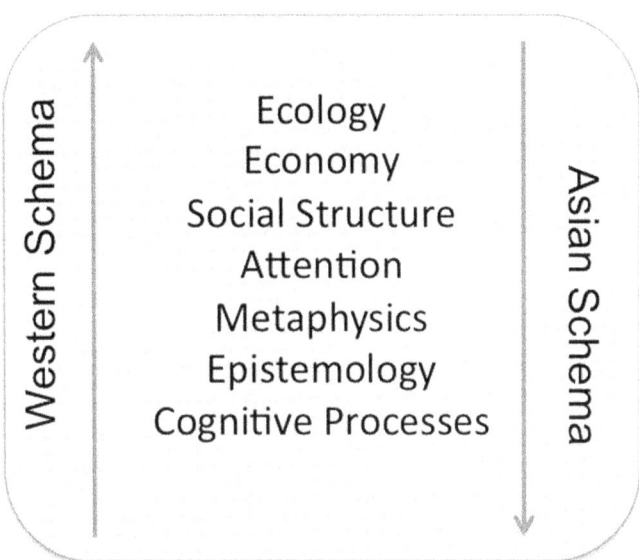

Figure 3. Western and Asian cognitive schema. Nisbett (2004) proposes that Westerners view the world first from the perspective of their cognitive processes, whereas Asians view the world from within their environment.

- Beliefs about controllability of the environment, with Westerners believing in controllability more than Easterners.
- Tacit assumptions about stability vs. change, with Westerners seeing stability where Easterners see change.
- Preferred patterns of explanation for events, with Westerners focusing on objects and Easterners casting a broader net to include the environment.
- Habits of organizing the world, with Westerners preferring categories and Easterners being more likely to emphasize relationships.
- Use of formal logical rules, with Westerners being more inclined to use logical rules to understand events than Easterners.
- Application of dialectical approaches, with Easterners being more inclined to seek the Middle Way when confronted with apparent contradiction and Westerners being more inclined to insist on the correctness of one belief vs. another. (Nisbett, pp. 44–46)

I want to guard against the presumption that culture automatically determines one's mental models, but rather to state that cultures have dominant mental models. Nonetheless, as Henrich, Heine, and Norenzayan (2010) have established in their research of human psychology experiments, the majority of research subjects at the basis of most psychological and cognitive theories are "W.E.I.R.D." (Western, educated, industrialized, rich, and democratic) and do not accurately represent the global population. Many of the assumed generalizations about the nature of human cognition are as skewed as studying hawks and then making generalizations about all birds. As such, mechanism is closely tied to the patterns of Western thought and should not be universalized.

In terms of media literacy, as shown in the research I'll discuss in Chapters 4 and 5, common methods stress the visual sense and a mechanistic model of communication and cognition. I believe this kind of response to media works at the level of events, patterns, and trends (as described in the iceberg model in Figure 2), and rarely, if at all, goes to the level of design and mental models. For media literacy to shift towards a more sustainable model of cognition and communication, I propose that it will be necessary to develop other methods that enable learners to not just view objects and the environment as embedded in their cognitive process but to see media and communication as embedded within a broader ecology. The discussion of ecomedia literacy in Chapter 6 explores how this can be achieved.

## The Metaphor Is the Message

Greening media education means transitioning to a *complementary relationship* with living systems by introducing ecological intelligence as balancing feedback in media literacy education's model of learning. One approach is to study how our language informs and shapes meaning systems. Bowers (2012) theorized that by reappraising root metaphors and changing how we use language, we can transform people's frameworks, and hence mental models. As such, in the following sections I examine the legacy of metaphor usage in the media disciplines with the aim of demonstrating that understanding metaphors enables us to explore and raise awareness of how mental models influence the way media literacy educators perceive media and the environment.

The study of media is so multidimensional that throughout the history of media studies scholars and critics have deployed a variety of metaphors to be used as "complex narratives about particular issues" (Jensen, 2002b, p. 248). As such, identifying these metaphors allows us to see how language structures ideologies and wider theory (Z. Todd & Harrison, 2010); metaphor usage offers an important lens for understanding the epistemological framework of a particular media discipline's perspective. Subsequently, the use of ecological metaphors in media studies is of particular interest, but it is also strange. For example, why do media studies practitioners use the ecology metaphor for the *problem* of media, yet exclude living systems as part of the discussion? Sontag (2002), who was concerned with how photos desensitize us to war, hoped for an *ecology of images* that would enable people to better understand how they impact us. Boyle (1997), an intellectual property scholar, envisions *information environ-mentalism* that preserves a commons of ideas from nefarious copyright enclosure. In response to violence in the media that creates a "mean world syndrome," Gerbner (1998) argued for a *culture environment movement*. McLuhan inspired Postman's concept of *media ecology* by promoting the idea that communications technologies create *media environments* (Logan, 2007). Lasn (2000), who founded the culture-jammer (media activist) magazine *Adbusters*, is concerned with preserving the *mental environment*.

Fuller (2005, p. 4) suggests that these perspectives have in common a kind of media *environmentalism*:

> Here, "ecology" is more usually replaced with the term "environment" or is used as a cognate term where the fundamental difference between the two concepts is glossed over. Echoing differences in life sciences and in various Green political movements, "environmentalism" possesses a sustaining vision of the human and wants to make the world safe for it. Such environmentalism also often suggests that there has passed, or that there will be reached, a state of equilibrium: that there is a resilient and harmonic balance to be achieved with some ingenious and beneficent mix of media.

Though the very words *environment* and *ecology* invoke a common good, ultimately these different uses are confusing and lack an underlying meaning. As Jennifer Daryl Slack (2005, p. 108) observes,

> Since the 1960s, the popular deployment, proliferation, and promiscuity of uses of "environment" and "ecology" exceed attempts to assign them identifiable referents. Infused with affect, they allude to "what matters," intimate something "critical" demanding attention, imply the importance of certain kinds of (inter)relationships, and invoke the idea of (re)connecting in ways that suggest much at stake. One cannot read a specific epistemology or politics off their use in general, even though their uses tend to carry some residue of their etymology.

Consequently, with the exception of the field of environmental communication, until recently the ecological concerns of past media polemics generally have not extended past metaphorical usage: *ecology* has been colonized by the rhetoric of media criticism, but in practice the principles of ecological understanding have not been adapted. Indeed, as noted by Heise's (2002) study of ecology metaphors in media scholarship, the invocation of *environment* in communication and information theory is devoid of reference to living systems. Heise argues that *media ecologies* are characterized by

> first, the way in which such technologies form a cultural environment that most of its inhabitants take for granted, but that nevertheless shapes their cognitive possibilities and social behavior in significant ways; second, the ways in which changes in one

individual technology change the media configuration and its manner of operation as a whole; and third, the ways in which such technologies function as systems with a logic of their own. (p. 157)

She is concerned that "When media are portrayed as cybernetic, self-regulating, and self-perpetuating systems...they are made to seem independent of political, social, and cultural interests and organizational patterns" (p. 157). As we will see, the misappropriation of ecology in media theory results from a map/territory disconnection in which ecology is not correlated with actual living systems. Borrowing from the language of semiotics and postmodern theory, in media studies, ecology has become an *empty signifier*.

## Media Metaphors

A number of conflicts among media literacy educators stem from their disparate views of media, which can be traced to their use of root metaphors to describe how media function. For example, some educators use medical terms to describe media as a disease that requires the inoculation of media literacy. Those who view media as grammar and language reject the medical approach and advocate for fluency and translation. Meyrowitz (1998, p. 9) points out that among media literacy educators

> there is less consensus about what we mean by media than many researchers, parents, and teachers may at first glance imagine... [D]ifferent ways of thinking about media lead to different conceptions of the competencies, or literacies, that may be desirable in the educated and aware citizen.

Meyrowitz (1980) argues that a specific metaphor adapted by communication media researchers can build entire theoretical frameworks because

- it helps define the key "issues" and "problems"
- it shapes the type of research questions that are asked
- it defines the type of data that is searched out (who do you interview, students or teachers?)
- it shapes the language in which the problem and results are expressed
- it determines to some extent the procedures that are used to examine and collect data
- it determines what problems, questions, data, and procedures are IGNORED. (Meyrowitz , 1980, p. 5, emphasis original)

Stemming from this approach, Meyrowitz identifies three metaphor classes commonly used to guide media research that act as a kind of methodology: *conveyer belt* (delivery of information objects from one place to another);

*language* (the grammar of visual codes such as camera angles and edits); and *environment* (a medium context). It is worth noting that Meyrowitz's use of the term *environment* omits living systems from its definition and is similar to how environment is used by other media scholars.

## Container Metaphors

As noted, metaphors are used because they simplify and ground complexity in experience that is familiar to us. This is what Lakoff and Johnson (1980) mean by metaphors being embodied: we associate *up* with *good* because in our bodies, being erect is associated with feeling positive. Likewise container metaphors are common in Western culture because we tend to differentiate between an *inner* and *outer* experience based on our perception of a clear separation between inside and outside of our bodies. Not surprisingly, container metaphors, like buildings, are often used to describe ideas: we use *frameworks* and *scaffolding* to *construct* our beliefs. From on ecocritical perspective, Garrard (2004, p. 10) insists the construction metaphor is problematic because it is too simplistic to assume that an "artifact like a building or machine" is an "autonomous work of minds and hands":

> I doubt many readers would automatically imagine a natural construction such as a termite mound. But if any building or machine, however technologically advanced, must be made by evolved animals (*Homo sapiens*) of materials of natural origin in accordance with natural "laws" of mechanical physics, then it follows that all our vaunted constructions are, in a sense, natural constructions. Perhaps the architectural metaphor obscures, or mystifies, the natural basis of all human culture and exalts only our own powers as a species. The excessively culturalistic implications of "construction" are not easily avoided by a substitution of terms, but I tend to use "shaping," "elaboration," or "inflection" to describe the complex transformations and negotiations between nature and culture or between real and imagined versions of nature.

Capra (2008) stresses that building metaphors for knowledge are used prominently in science, which depends on theoretical *foundations* and *basic building blocks*, but these terms are inadequate for systems approaches. Instead, Capra asserts,

> In the new paradigm, the metaphor for knowledge as a building is being replaced by that of the network. Since we perceive reality as a network of relationships, our descriptions, too, form an interconnected network of concepts and models in which there are no foundations. (2008, p. 368)

Given that media scholars and media literacy practitioners come from similar *language communities* as social science, it is not surprising that container metaphors are as pervasive as they are in media studies and media education.

## Conveyer Belt and Transmission

Meyrowitz (1980, p. 7) contends that with the conveyor-belt metaphor, "The medium is seen as a passive delivery system of important messages." The methodology derived from this view is *content analysis*, which is a *medium-free* approach to media texts that focuses on particular messages (such as surveying acts of violence in television programming). The conveyor-belt metaphor parallels Carey's (2009) explication of the *transmission* metaphor that dominates traditional communication theory. In this view of communication, *transmission* reifies ideas as objects passed through space. For Carey, transmission

> is defined by terms such as "imparting," "sending," "transmitting," or "giving information to others." It is formed from a metaphor of geography or transportation. In the nineteenth century but to a lesser extent today, the movement of goods or people and the movement of information were seen as essentially identical processes and both were described by the common noun "communication." The center of this idea of communication is the transmission of signals or messages over distance for the purpose of control. (2009, p. 12)

This dovetails with Reddy's (1979) critique of the *conduit* metaphor, which is an assumption that words are like boxes that can be filled with meaning that then can be passed between people. The everyday use of this metaphor ("Who *gave* you that idea?"; "Do you *get* what I'm saying?"; "I'm trying to *get* this idea *across* to you"; etc.) suggests that we internalize the notion that communication is without context or local meaning. Bowers (2009, p. 34) calls the conduit metaphor a "minor myth" that "has huge ecological and cultural consequences":

> It is necessary to the support of three other minor myths: namely that the individual is an autonomous thinker (at least has the potential to be), that there is such a thing as objective data and information, and that the rational process transcends all forms of cultural influence. The conduit view of language, which also is reinforced in print-based modes of communication and storage (which includes computer-mediated thinking and communication) leads to thinking of words as having a universal and timeless meaning that transcends cultures. It also contributes to ignoring that abstract words marginalize awareness of local contexts, tacit understandings, and embodied/culturally mediated experiences.

The *conveyor belt*, *transmission*, and *conduit* metaphors are examples of a unidirectional model of communication and are implicit in early communication theory. For example, the linear view of communication referred to as the *Shannon-Weaver model* is associated with the *magic bullet* or *syringe* metaphors, the idea being that media penetrate our minds like a bullet or syringe. This reinforces Enlightenment notions of how the autonomous individual's mind is a machine-like processor of representational images that move through space. In this model, a sender codes a *message* that is unidirectionally delivered through a channel to be decoded by a receiver. Peters' (1999, p. 23) study of the history of the *idea* of communication suggests that to Shannon, "'communication theory' was explicitly a theory of 'signals' and not of 'significance.'" Likewise, Peters argues (1999, pp. 63-64), the modern idea of communication is rooted in concepts of key European philosophers, whose ideas are foundational to modernity and mechanism:

> Augustine and Locke both provide articulate defenses (but with very different purposes) of ideas fundamental to the modern notion of communication: the interiority of the self and the sign as an empty vessel to be filled with ideational content. In its everyday usage, "communication" rests squarely on such conceptions: Each of us has a treasure chest of thoughts and wishes uniquely our own. Our interiors are private, goes the tale, and trapped inside by the privacy of our senses and the individuality of our minds.

As Meyrowitz (1980, p. 6) proposes, "Most media studies are based on this image of media as a 'conveyor belt,'" and therefore, are taken for granted. For example, *The Penguin Dictionary of Media Studies* (Abercrombie & Longhurst, 2007) defines communication as "The transfer of MESSAGES from one party to another" (p. 69). In a separate entry, the authors state that a message is, "A unit of communication between a sender and receiver" (p. 223). This standard view is incorporated into the generic communication formula of mass-media studies, which frames any communicative situation as, "who says what, in which channel, to whom, with what effect" (Morley, 2005a, p. 49). Such a communication model, Carey (2009, pp. 26-27) argues, becomes tautological:

> The widespread social interest in communication derives from a derangement in our models of communication and community. This derangement derives, in turn, from an obsessive commitment to a transmission view of communication and the derivative representation of communication in complementary models of power and anxiety. As a result, when we think about society, we are almost always coerced by our traditions into seeing it as a network of power, administration, decision, and control— as a political order. Alternatively, we have seen society essentially as relations of property, production, and trade—an economic order. But social life is more than power and trade (and it is more than therapy as well). As Williams has argued, it also

includes the sharing of aesthetic experience, religious ideas, personal values and sentiments, and intellectual notions—a ritual order.

As my research indicates, container metaphors are pervasive in media literacy education. For example, Potter (2004) builds his pedagogical framework based on a view of media as the "technological dissemination of messages" from one entity to another, messages being "those instruments that deliver information to us" (pp. 43-44). This language use replicates an underlying cultural assumption that knowledge is disembodied and contextless, which invariably reinforces mechanistic assumptions about the individual as an autonomous being independent of the environment. As Bowers (2009) argues forcefully, such a worldview underlies unsustainable cultural practices.

An example of how this plays out in media literacy is as follows. The traditional media literacy approach uses media deconstruction as a primary pedagogical technique. This means that learners analyze specific texts in an effort to unlock latent, hidden, or concrete messages that have been coded by mediamakers. This method assumes that a specific message is delivered through the media text and must be unpacked by the learner. By reifying information, this method reinforces the notion that media producers and consumers are autonomous from their respective environments, and that the messages lack a history, context, or relationship outside the specific reality of the media text. Such an approach reinforces a top-down linear communication model in which producers (*programmers*) and consumers (audiences that are *programmed*) are functioning within a machine-like communication system. From an embodied cognition perspective, a media message is more similar to an intention that sets in motion a series of responses that vary according to social and physiological environments. For example, consider how one of the most highly mediated events of our lifetime, the terrorist attacks of September 11, has been so widely interpreted, setting in motion a diversity of responses. As such, learners are situated within particular ecologies, which explains why a 14-year-old from southern California will respond differently to a McDonald's ad than will a 40-year-old Zapotec mother in Oaxaca, Mexico.

## Language and Grammar Metaphors

Meyrowitz (1980) also observed another class of media metaphors, language and grammar, which entails the grammar of media construction, such as camera angles, edits, or sound. This approach to media is primarily the province of film studies. These *production variables* are more format- and context-sensitive (i.e., they vary according to medium such as film versus radio), but I believe they are vulnerable to a mechanistic methodology if treated ethnocentrically. That is, there is a danger of media language being treated as universal in the same way that Hollywood blockbusters are viewed as a standard model for films. Visual grammar needs to be approached as constructed within the history of Western culture in the same way that Berger (1973) argues "ways of seeing" are culturally conditioned. For example, when we see a movie or TV show, our expectations of how their visual grammar are deployed and make sense are the result of our vision being constructed through our exposure to linear perspective in art, photography, and moving images, which have an implied verisimilitude. We assume that film and video continuity editing techniques are natural, yet they don't make sense to people who have not been exposed to either medium. Furthermore, as feminist media critics have argued, we are also conditioned to see women through a "male" gaze (Mulvey, 2001). The objectification of male and female bodies in media is as much a result of editing and camera angles as culture and economics. Due to a bias of focusing on message analysis, language and grammar metaphors are less likely to be encountered in media literacy.

## Environment Metaphors

The third media metaphor class discussed by Meyrowitz, *environment*, is specifically related to medium theory, which "emphasizes the importance of media technologies in determining the features of media products and content, as well as determining their social, cultural, political and economic uses" (Laughey, 2007, p. 202). According to this perspective, various media such as radio, film, TV, print, and internet have different ways of communicating and creating contexts. Each has its own particular meaning design. Here,

> The medium is seen as a type of social context or social situation that includes and excludes participants. And like most environments, the people who have access to it share an experience that gives them a sense of group identity, while those who are excluded from the environment are also excluded from the sense of belonging. (Meyrowitz, 1980, p. 11)

This area is the most useful for a green-media education approach because it is systems oriented. For example, if we look at commercial media produced as part of the project of globalization, then we can see that the tautology of technological progress embedded in media technologies excludes bioculturally diverse perspectives and are, in Shiva's (1993) terms, *monocultural*.

Medium theory and the view of media as environments are closely associated with media ecology. Scholars in the media ecology tradition often use the term *ecology* according to a generic definition in which it represents a *system of systems*, as opposed to the understanding of ecology as the study of biological systems. Though the field's notion of ecology is somewhat mechanical in the same way that systems theory can also be reductive, media ecology's methodology forges an important environmental link through an interdisciplinary approach that goes beyond the social sciences and humanities to include cognitive and other sciences. Lum's (2006) history of media ecology reveals that the expanding discourse of environmentalism had propelled an interdisciplinary turn in the 1960s and 1970s, thereby creating the context in which the core ideas of media ecology could emerge:

> What is significant to note has been the fact that the environmental consequences brought forth by rapid technological advances in the early part of the 20th century have fostered an ecological paradigm or way of thinking about the interconnectedness of things in people's lives…. It is within this larger intellectual context at the dawning of the "ecological age" that we see the significance of the convergence of interests from among diverse academic orientations in understanding the foundational and ecological impact of technology, a defining issue of media ecology's paradigm content. (Lum, 2006, p. 16)

This context helps explain Logan's (2007, p. 3) view of how *ecology* became the field's primary metaphor for studying media:

> The introduction of the term ecology into what had been called media studies or communication studies signaled the fact that the study of media by media ecologists was not merely a study of the content of media. Rather, media ecology entails a study of the social, cultural and psychic impacts of media independent of their content embracing McLuhan's defining one-liner: the medium is the message.

Again, note the conspicuous absence of natural systems in this specific reference. It certainly is in the background and of concern to many in the field, but most works associated with media ecology are absent an explicit green discourse that foregrounds the discussions emanating from environmental philosophies or those promoted by environmental communication.

Media ecology's primary concern is the way media serve as sensory extensions and favor particular cognitive biases, such as how oral communications correspond with the ear, and print with the eye (Ong, 1982). By focusing on medium theory, this field addresses the degree of technological determinism and cognitive bias inherent in media systems (print, telegraph,

TV, internet, etc.), and to what extent they shape cognitive environments, ranging from *soft* to *hard* determinism (Lum, 2006, p. 34). For the purpose of ecomedia literacy, the benefit of the media ecology approach is that it stresses medium and technology as powerful forces central to social practice. Its weakness is a lack of discourse surrounding the natural-world aspects of ecology, and the downplaying of how the mythical dimension of media (which is highlighted by content analysis) shapes cultural beliefs (McLuhan [2002b, p. 18] thought content was like meat for the guard dogs, and wasn't important for the study of media). More useful for the purpose of ecomedia literacy is Fuller's (2005, p. 22) concept of media ecology, which is "where media elements possess ontogenic capacities as well as being constitutively embedded in particular contexts." Such a definition can apply as much to corporate TV as to pirate radio. The benefit of Fuller's model is that it acknowledges how local contexts can temper the mechanically deterministic approach advocated by past media ecologists.

## (Re)Mediating Media Metaphors

Bowers (2012, p. 107) suggests reframing metaphors with new analogs in order to update them to match the challenges of sustainability. In particular,

> The recent emergence of ecology as a root metaphor provides an interpretative framework that brings into question the root metaphors of individualism, progress as a linear form of change, and an anthropocentric universe. It also brings into question different interpretations of evolution that emphasize competition between organisms in meeting the test of Darwinian fitness—and that cultural memes must meet the same test. As an interpretative framework, ecology emphasizes interdependence within systems that are nested in even more complex systems. Unlike the root metaphor of evolution, it cannot be used to justify political and economic ideologies such as libertarianism and market liberalism.

Reframing media as an ecosystem is an important step in this direction, but as the previous discussion showed, the ecosystem metaphor can have many meanings depending on the context and how it is used. In some interpretive frameworks the metaphor can even work against the principles of sustainability and ecology. I suggest one way of reclaiming media ecosystems as embedded within living systems is to use the term *biocultural media ecosystem*. Bioculture refers to how culture is intrinsically connected to ecological identity. This rephrasing is an example of what I call *(re)mediation*. The term remediation means the repair of a damaged ecosystem; mediation is the manner in which the world is mediated. The combination of these terms, (re)mediation, means repairing the media metaphors that have up to now depended on mechanistic approaches to communication and cognition.

In a biocultural media ecosystem, we shift from a communication theory based on industrialism to one scaled to human ecosystems in which information and communication interact organically between local and larger living systems. This approach is supported by Carey's (2009, p. 19) *ritual* approach to communication. Carey asserts, "communication is a symbolic process whereby reality is produced, maintained, repaired, and transformed." We can interact with these processes more organically by incorporating ritual aspects of communication that were obscured by the Industrial Revolution's transmission model:

> In a ritual definition, communication is linked to terms such as "sharing," "participation," "association," "fellowship," and "the possession of a common faith." This definition exploits the ancient identity and common roots of the terms "commonness," "communion," "community," and "communication." A ritual view of communication is directed not toward the extension of messages in space but toward the maintenance of society in time; not the act of imparting information but the representation of shared belief.... If the archetypal case of communication under a transmission view is the extension of messages across geography for the purpose of control, the archetypal case under a ritual view is the sacred ceremony that draws persons together in fellowship and commonality. (Carey, 2009, p. 15)

The ritual approach of communication complements Bowers' (2008) concept of the cultural commons, which is an intergenerational cultural space based on shared knowledge, a kind of self-correcting system that could ensure sustainable cultural practices.

In addition, given that *message* is the primary unit of analysis in media literacy and that it is clearly tied to mechanism, it is necessary to adapt an alternative concept of information. One way is to approach media texts as *intertextual*, which means that texts are shaped by other elements outside of themselves, such as genre, discursive conventions, discursive communities, and cultural codes. The text does not contain communication objects as the conveyer-belt metaphor suggests; rather meaning is determined by its larger thinking system. Gray (2006) asserts that media texts (such as media literacy education documents) work through and against each other; texts only exist in context of other texts, and are always becoming. He draws on Bakhtin and Holquist's (1981) theory of *dialogism*, which was a reaction to Saussure's linguistic structuralism. Structuralism is the belief that all language has an underlying, unifying grammar that structures the way people communicate. Instead, Bakhtin and Holquist believed that communication is based on an utterance: there is no beginning and end of communication, but an ongoing *chain*. In this model, "The text is always in flux" and is a *field of action* (Gray, 2006, p. 29). This correlates with DeLuca's (1999) discussion of image events (such as direct actions staged by environmental groups to draw media attention), which take place within a diverse *heteroglossic public sphere* where meaning

> happens at the site of the audience, but it is marked not by unity, integrity, faithfulness, and finality, but by conflict, contradiction, complexity, and contingency, the result of negotiations between audiences, texts, authors, and contexts where none of these constituent elements is self-identified or originary... there is no site that collects the irreducible multiplicity of meanings. (DeLuca, 1999, p. 145)

This is in stark contrast to *Western metaphysics*, a model that adheres to "hierarchical binary oppositions" like "transmission/dissemination, presence/absence, immediacy/mediation, speech/writing, author/audience, text/audience, communication/miscommunication, reason/emotion, culture/nature, human/animal" (DeLuca, 1999, p. 140).

The reader of these "intertexts" is not merely an autonomous mind, but perceives the world through embodied cognition. As discussed, this concept of thinking is situated within self-organizing systems constrained by environmental affordance (Maturana & Varela, 1998, 1999; Varela, Thompson, & Rosch, 1991). So, like Nisbett's (2004) discussion of the difference between Western and Asian schemas, embodied cognition means the mind is embedded in living systems (as opposed to nature being an object that the mind observes). This differs from mechanism because people are understood to be part of thinking systems that extend beyond their bodies into the spaces they embody, including regional habitats and computer networks. Unlike mechanism, from the perspective of embodied cognition, communication is a *disturbance* with no fixed outcome other than what the environmental context affords. A disturbance, to borrow from Gregory Bateson's (2000) terminology, is a "difference that makes a difference." I am assuming that people who view themselves as embedded within living systems will care more about the health of those systems, thus embodied cognition is fundamental for advancing sustainable mental models.

## Monoculture and Permaculture

In order to green media education we can also reclaim some media terms to recontextualize them within an ecological intelligence framework. For instance, the term *broadcast* originates from the practice of casting seeds into a field (Wu, 2010). A field is also used to describe an area of reception (as in a signal's field). Culture derives from agriculture and was originally meant to describe a process of cultivation. In the most profound but simplest of terms, media grow culture. Not only that, we can take this a step further by acknowledging the brain is a living system, and that cognition is something that can be cultivated. Indeed, ecopsychologist Craig Chalquist (2010) reminds us that patterns repeat themselves in nature (such as spirals and branching systems) and that the brain and its stem can be likened to a tree. He points out that many of the words we use for brain anatomy have etymological roots in nature, such as amygdala ("almond"), pineal ("pinecone"), cortex ("bark of a tree"), and dendrite ("tree"). Even the word *human* comes from "humus." Imagine treating minds not as machines that need programming, but as gardens or forests that need tending and care.

Along these lines, Gadotti (2010, p. 208) suggests that sustainability education (and by extension, green media education) can be conceptualized with the metaphor of gardening:

> The garden allows for working with the Earth, learning to care for the 'fabric of life' (Capra 1996); perceiving the Earth through the Earth; seeing the seed assume the form of the plant and the plant assume the form of food, the food that gives us life. It teaches us patience and careful handling of the Earth between sowing and harvesting. In gardening, we learn that things are not born ready made; that they need to be cultivated and cared for. We also learn that the world is not ready made, it is being made, it is making us; that building it demands persistence, hopeful patience of the seed, which at some moment will sprout and flower, and will be fruit.

Subsequently, there are different philosophies about how to grow, such as *monoculture* and *permaculture*. Shiva (1993) writes about how monoculture is not just an agricultural activity, but also a culturally specific philosophy of economics, education, and social structure that is universalized through the global economic system. The agricultural practice of permaculture is a different heuristic because it is a systems-oriented approach based on the relations and connectivity of various elements that go into cultivation in a local environment. Holmgren (2002, p. xix) defines permaculture as

> Consciously designed landscapes which mimic the patterns and relationships found in nature, while yielding an abundance of food, fibre and energy for provision of local needs. People, their buildings and the ways they organise themselves are central to permaculture. Thus the permaculture vision of permanent (sustainable) agriculture has evolved to one of permanent (sustainable) culture.

When applied as *media permaculture*, such an approach incorporates the concepts of appropriately scaled communication technology as proposed by

Illich (1973) and a kind of "farmer's market" of diverse media suggested by McKibben (2008). Comparing and contrasting media within the parameters of these agricultural metaphors can frame media ecosystems as either biocultural or mechanistic. Furthermore, if texts are viewed as nodes within networks of diverse worldviews, learners begin to think more systematically about relationships and connectivity rather than in terms of reductionism and atomization.

Finally, *ecosystem* should repurpose *oikos* (household) as a dwelling metaphor. This approach creates a space for understanding media ecosystems as constituting habitude and habitat. Consider how households are primary sites of mediation. For example, it is where we often work on the computer, watch TV, use our phones, share meals, and socialize, all being mediated through language, metaphors, and economic practices. Economic globalization is intimately integrated into our homes through the goods we consume, the power we use, the food we eat, and the culture we share and experience. Within an ecological intelligence framework, media education then becomes a process of understanding the permeated, networked border zone between our own cultural environments and those of larger biocultural media ecosystems.

CHAPTER THREE

# A Field Walk Through the Media Ecosystem

Len Masterman's (1989) pivotal book, *Teaching the Media*, began by asking, why teach media? Masterman explained that while media had become "increasingly central components of social, economic, and political activity at all levels, media education [remained] marginal within educational systems everywhere" (p. 1). Subsequent efforts (Lusted, 1991; Tyner, 1998) were intended to expand the possibilities and connections for media education across disciplines in educational settings that traditionally privilege print literacy over the inclusion of electronic media as an area of inquiry. These early arguments for media education reflected debates emerging from media studies, cultural studies, film studies, and literacy scholarship. Unlike their academic counterparts who asked how media impacts society and culture, the early pioneers of media education were concerned with pedagogy: What are the best practices for teaching media? This essential question is based on an assumption that media education, like print literacy, is integral for promoting life skills, social capital, and cultural citizenship. The initial aim of media education, as outlined by Masterman and the many practitioners who followed his lead, recognized the power of media to shape attitudes and beliefs, and believed that education should address our engagement with media.

In terms of understanding the worldview of media literacy educators, Masterman's work highlights two themes. First, his discussion of media education comes from the perspective of media studies, film studies, and cultural studies. As it turns out, media literacy intersects and converges with many different perspectives and fields, such as literacy, public health, education, media arts, and psychology, yet media studies remains one of its strongest influences. Given the way knowledge is siloed in academia, it is very rare to find practitioners who have mastered all fields. With health workers, community activists, mediamakers, teachers, academics, media professionals, and artists participating in the field of media literacy, each offers a particular strength and emphasis.

Second, Masterman wrote during a formative period for media literacy, at a time when it was solidifying as a movement. Though media education has roots going as far back as the 1920s, the majority of organizations analyzed during my research formed in the 1980s and 1990s. Indeed, the core definition for media literacy mentioned in the introductory chapter was formulated at the Aspen Institute in 1993. Noting the time period is important because the philosophies and pedagogical strategies that have become part of current practices and debates are challenged by an assortment of external factors, such as the emergence of social media, ubiquitous personal media gadgets, funding restrictions, education reform, and global economics. Indeed, Bennett et al. (2011) consider the internet as media education's "inconvenient truth." Even though the practitioners I interviewed are grappling with these changes, the primary documents that were researched reflect previous debates and formulations. For this reason, it is necessary to compare the current media ecosystem with media studies and media literacy debates to see how they are co-evolving.

## The Media Ecosystem

In everyday vernacular when people talk about both the positive and negative aspects of entertainment, popular culture, and news, *the media* (as a singular entity) usually signifies the mass-produced culture industry represented by communications technology such as television. As Morley (2005b, p. 212) illustrates,

> Nowadays the term "media" is most commonly used to refer to the institutions of electronic broadcasting, printing magazines, and newspapers which address mass audiences. In this context, by contrast to interpersonal or two-way forms of communication, the emphasis is usually on the sense in which mass media constitute powerful one-way systems for communication from the few to the many. In societies with advanced systems of division of labor, people tend to live highly segregated lives, and are thus increasingly dependent on the media for information about events outside their own immediate experience. To this extent, contemporary societies can be claimed to be characterized by the mediation of much of our social experience. Indeed, it has become common to refer to the contemporary world as comprised of mediated and mediatized societies.

This description is important because when media literacy practitioners refer to media, the term usually implies *mass media* while excluding other forms, such as participatory, alternative, and social media. Until the rise of the web in the early 1990s, mass media generally reflected the state of communications in industrialized countries. This characterization of media has dominated media studies, which is a multidisciplinary field of research and theory about how media impact society, culture, and audiences. When it comes to exploring media studies, obviously the cultures of specific disciplines within the field have their own distinct approach, emphasis, and style that weigh how media are analyzed and taught. For example, humanities and social sciences often have divergent goals reflected in the split between qualitative and quantitative methodologies.

Though contemporary media studies is now institutionally globalized, the early pioneers of mass-media scholarship are identified with the Chicago and Columbia Schools, which were primarily sociological and positive in their view of the role of media in society. As mentioned, several scholars from the Chicago School, such as Robert E. Park and Harold D. Lasswell, applied concepts from human ecology to media, but never linked ecology with living systems. Critical theory, which has had a much larger influence on media studies, is associated with the Frankfurt School's neo-Marxist orientation that viewed media as having a negative, top-down influence. Media ecology came out of the Toronto School and is associated with medium theory. Cultural studies is connected with Birmingham's Center for Contemporary Cultural Studies; it was historically neo-Marxist and critical but parted from the Frankfurt School by incorporating ethnographic audience research and proposing a more dynamic interplay between audiences, culture, society, and industries. Film studies is more diffuse and a discipline unto itself, but it has also had an important influence on media literacy and media studies, in particular the British Film Institute, which produced some of the earliest critical viewing curricula. Finally, it is important to acknowledge a number of important areas of inquiry that inform media studies, including structuralism, semiotics, feminism, postcolonial theory, postmodernism, information society, and popular culture studies. What is important here is to examine the overall texture of how media as a whole are perceived.

By the time of media study's formative period from the 1920s to 1940s, the realms of inquiry and divisions of sciences already divided ecological approaches between either physical or social sciences, but never combined. More importantly, Boyd-Barrett (2002, p. 30) argues that early media theorists reflected the prevailing mass society paradigm of their time:

> In brief, the first major phase of media research was characterized by a mass society model of society, by a focus on the impact of media on the moral robustness of the community as a whole. It viewed the media as very powerful, and its model of the relationship between media and reader or consumers was a transmissional one,

sometimes described as the "hypodermic needle" model of media effect. Its tone was overwhelmingly negative in its appraisal of the role of at least the popular media. Prevailing methodology was deductive reasoning on the basis of evaluative premises of the nature of human beings and of their potential.

Though, in this period, scholars spoke of *mass culture, mass society,* and *mass communication,* the notion of *the media* coming to mean *mass media* wasn't common until the 1960s with the rise of television (Scannell, 2002, p. 194). In most cases, though, the implicit understanding was that media were distributed in the *one-to-many* form: a core group of centralized media producers broadcast their products to an anonymous audience with uniform effect.

Subsequently, media studies has roots in the cultural assumptions of the Industrial and Scientific Revolutions. Many of the taken-for-granted concepts derived from the early era remain strong currents in how people think about media today, including ideas about technological progress, human agency, autonomy, and a mechanistic model of the mind. For example, Walter Lippmann's 1922 book, *Public Opinion* (2007), proposes a mechanistic framework of communication and cognition. Lippmann believed that people formed opinions about the world based on the "picture in their heads," and it was the job of media, for better and for worse, to shape these pictures. Lippmann's views corresponded to the rise in the 1920s of behavioral psychology and advertising, both professions excited by the idea that human behavior could be modified through a kind of mechanistic conditioning and "manufactured consent." Subsequently, during the period of Lippmann's book, it was believed that the marriage of behavioral psychology and marketing would solve a problem for industry that needed to convince people to buy into the emerging culture of mass consumption (Ewen, 2001).

Lippmann's book contains key themes that are important to the discussion of media study's history. He is critical of media's power of influence, so he believes mass media should be carefully managed by chosen elites. This tension between a critical view, which was later taken up by the neo-Marxists of the Frankfurt School, and the *administrative* approach of the Chicago School, which views the positive potential for media to promote "democratic values" (and hence capitalism), represents the back-and-forth seesaw of mass-media studies up through the 1960s. The fulcrum that the two poles balanced on was a mechanistic, or injection, view of linear communication influenced by the Shannon-Weaver model. Though the aims of the Frankfurt School and the administrative approach are at odds with each other, they both take the position that media effects are largely linear: mass culture is programmed by mass media without regard to local contexts.

An alternative approach to the linear (top-down) model of critical theory and effects emerged from cultural studies, including the concept of the *circuit of culture* (Du Gay, Hall, Janes, & Mackay, 1997). It de-emphasizes political economy in favor of a balanced analysis that takes into consideration how culture and material production are recursive. Whereas the sociological approach of effects can stress how economic production orders culture, the circuit of culture model argues that culture is part of a complex feedback system. The interdependent elements of the circuit (representation, identity, production, consumption, and regulation) can provide a framework for studying any media text or gadget (e.g., the iPhone) and its multiple dimensions. Such an approach tempers the more deterministic strands of media ecology, which argues that media technology tends to be the root structuring mechanism of society, and counters the critical theory view that economies are the core structuring mechanism.

As it relates to ecomedia literacy, the key is to shift the circuit of culture's five processes into an ecological framework (such as looking at the environmental aspects of material production, consumption, and electronic waste, and challenging a mechanistic model of culture and identity) so that one could explore a cultural object such as the iPhone through the lens of green cultural citizenship. Such an approach would have an implicit ecological ethic that is less neutral than many of the cultural studies methodologies that tend to be uncritical of media consumption and overly celebratory of audience "empowerment."

## Media as Education

According to Dominick (2009), the functions of traditional mass media include informing us and interpreting events (a teacher function); performing surveillance (a watcher function); servicing the economic system (with information and advertising); transmitting values and holding society together (acting as a kind of cultural glue); providing entertainment (and pleasure); linking and acting as a community forum (the media equivalent of a town-hall meeting or group discussion); setting the agenda for what is important (a stage for the theater of action); and servicing the political system (a venue for debate, exposure, and discourse). Given that mass communications are largely designed and shaped by gigantic multinational corporations (Bagdikian, 2004; Herman & Chomsky, 2002; Kellner, 2003; McChesney, 1999; Parenti, 1986), critics argue that most media companies promote a kind of *hidden curriculum* that "demands" a worldview (O'Sullivan & Taylor, 2004, p. 6).

As such, traditional media such as TV serve an instructive purpose in daily life, in particular regarding our beliefs about the economy and consumption. According to Cortés (2005, p. 55),

> The mass media teach whether or not media makers intend to or realize it. And users learn from the media whether or not they try or are even aware of it. This means all of the media, including newspapers, magazines, movies, television, radio, and the new cyberspace media. Such media serve as informal yet omnipresent nonschool textbooks.

Crucially, mass media play an instructional role by defining the status quo, setting the agenda of our socio-economic system, defining what to think about, and recursively reinforcing non-sustainable cultural beliefs. I concur with critics who believe we should be concerned about how narrow economic interests have the capacity to shape culture through their powerful manipulation of electronic communication. In this sense, Giroux (1994, pp. 29-30) asserts that corporate media inform *cultural politics* and *social practice*, and therefore perform a pedagogical function that leads to

> production of and complex relationships among knowledge, texts, desire, and identity; it signals how questions of audience, voice, power, and evaluation actively work to construct particular relations between teachers and students, institutions and society, and classrooms and communities.

Sterling's (2004, p. 32) critique of education makes a similar case: "in an age of mass communications, the socio-cultural milieu arguably affects people and influences values more than formal education programmes do." Subsequently, one of the primary by-products of media is learning how things work. As Thomas (1995, p. 447) maintains,

> 'educational' material coming from television stories probably has more to do with the business of ordinary life-values and ideas involved in our everyday judgments than does the educational material in most formal classroom situations...we may be more receptive to TV information than to that from classroom lectures precisely because we do not view TV stories as education. Indeed, it is unlikely that we would acknowledge or even recognize the source of much TV-gained information because its entrance into our store of knowledge is so subtle.

## Media Evolution

The meaning design that characterized mass media is now rapidly evolving. During the era dominated by one-to-many mass media, it was easier to distinguish between the producers and consumers of information, and also between the "good guys" and "bad guys" of media production. Throughout the history of media, as new media technologies emerged (such as telephony, film, radio, and television), the openness of media systems quickly gave way to closed monopolies (Wu, 2010). Predominantly there were relatively small groups of media companies producing the majority of media content, which was distributed in a linear (top-down) fashion with very few feedback mechanisms. The concept of *independent* and *critical* media was clearer and easier to distinguish because marketing and corporate media had yet to incorporate the aesthetics, practices, and politics of alternative media into their offerings. With the advent of postmodern media practices and the internet, these boundaries are now less clear. Corporate monopolies are much larger, but aesthetically they have incorporated avant-garde aesthetics and the language of irony so as to co-opt the traditional tools of critical media. This includes the co-optation of environmentalism and sustainability rhetoric, which enables media corporations to greenwash their organizations and practices. For example, News Corporation can claim to be carbon neutral, yet Fox News, its flagship network in the United States, continues to be a platform for anti-environmental ideology (Maxwell & Miller, 2012).

Nonetheless, as in the early days of telephony, radio, and film, with the rise of the internet, media's distribution and feedback structure has changed so that many more people can actively participate in the creation of media. As noted by Shirky (2010, p. 54, emphasis original),

> The bundle of concepts tied to the word *media* is unraveling. We need a new conception for the word, one that dispenses with the connotations of "something produced by professionals for consumption by amateurs".... Here's mine: media is the connective tissue of society.

According to Shirky, the old distinction was between two kinds of media—public (broadcast) and private (personal communications such as phone conversations or letter writing)—but now the two have fused.

Whether we are talking about industrial-scaled media, or *few-to-few* communications of social media, it is important to consider the pedagogical dimensions of our communications infrastructure. For those of us in rich, industrialized countries, computer networks are embedded in our lives as much as the places where we live; our cultural practices are deeply impacted by mobile communication gadgets (Moores, 2012). For example, in 2013, the Pew Research Center surveyed a nationally representative sample of parents and youth ages 12-17 in the United States to find:

- 78% of teens now have a cell phone, and almost half (47%) of them own smartphones. That translates into 37% of all teens who have smartphones, up from just 23% in 2011.
- 23% of teens have a tablet computer, a level comparable to the general adult population.
- 95% of teens use the internet.
- 93% of teens have a computer or have access to one at home. Seven in ten (71%) teens with home computer access say the laptop or desktop they use most often is one they share with other family members. (Madden, Lenhart, Duggan, Cortesi, & Gasser, 2013, para. 2)

Thus, we are increasingly a mobile, networked society with our cultural practices deeply impacted by technological design (Castells, 2011; Rushkoff, 2011). As such, in a general sense, "production today has to be conceived not merely in economic terms but more generally as social production—not merely the production of material goods but also the production of communications, relationships, and forms of life" (Hardt & Negri, 2004, p. xv). So when Jensen (2002a, p. 9) argues that media are *institutions-to-think-with*, what studying media offers us is an opportunity to map cultural paradigms because media are central to "the production and circulation of meaning in modern societies, enabling collective reflexivity and coordinated action on an unprecedented scale." In this sense, *media systems are forms of meaning design*.

In response to the changing media paradigm, Gauntlett (2007) coined *media studies 2.0* to suggest that media studies needs to evolve beyond the mass-media model to incorporate the media practices of web 2.0 (also called the *social web*). Web 2.0 includes social networks (Facebook, Twitter, YouTube), collective intelligence (Wikipedia), mash-ups (remixing), and participatory media (blogs). Media textbooks often present new media as specific categories different than books, television, or film, while continuing to offer the mass-media model as the primary lens for studying media (e.g., Baran, 2004; Campbell, 2009; DeFleur & Dennis, 2002; Dominick, 2009). Unfortunately the mass-media (one-to-many) model of analysis is not sufficient in an environment that is more characterized by networking (many-to-many and few-to-few) and convergence (the integration of media content across platforms and gadgets). New media practices complicate old notions of media production and consumption because the old model distinguishes between audiences and producers, whereas with new media users are an ambiguous hybrid of producer and consumer (*prosumer*) who are interacting more actively with media. Emerging practices include *convergence culture* and *fan culture* (Jenkins, 2006), social media and the capacity for people to self-organize more easily (Shirky, 2008), *remix culture* and the Creative Commons' alternative to traditional copyright (Lessig, 2008), creative digital storytelling across media (Alexander, 2011), and a *networked economy* that is fundamentally challenging top-down capitalism (Benkler, 2006). Additionally, in the rich industrialized world, we live in an environment of pervasive computing, which means that we inhabit a world of networked objects with chips: "Instead of pulling us through the looking glass into some sterile, luminous world, digital technology now pours out beyond the screen, into our messy places, under our laws of physics; it is built into our rooms, embedded in our props and devices—everywhere" (McCullough, 2004, p. 9). Thus, "The world of computing has changed. Information technology has become ambient social infrastructure. This allies it with architecture. No longer just made of objects, computing now consists of situations" (McCullough, 2004, p. 21).

The evolving situation of media and information technology offers new affordances, potentially encouraging increased participation and civic engagement. As some assert, we are in an emerging *participatory culture* environment

1. With relatively low barriers to artistic expression and civic engagement
2. With strong support for creating and sharing one's creations with others
3. With some type of informal mentorship whereby what is known by the most experienced is passed along to novices
4. Where members believe that their contributions matter
5. Where members feel some degree of social connection with one another (at the least they care what other people think about what they have created). (Jenkins, Puroshotma, Clinton, Weigel, & Robison, 2005)

Participatory culture can be viewed positively, but it also needs to be problematized when its tools are also instruments of profit for large media platforms such as Google and Facebook. Critics contend that the internet is largely governed by the rule of thumb that if the platform is free, then the user becomes a product that is sold to marketers; we turn ourselves into commodities by reifying our consciousness for private companies that own internet sites (Lanier, 2010). This complicates the notion of civic engagement because the sphere of engagement is increasingly uncertain and privatized. Given these current conditions (increased monopolization and privatization of media) coinciding with the rise of participatory culture, how can media practice lead to green cultural citizenship?

First of all, if we see ourselves as members of a media ecosystem, we are both cultural workers and organic intellectuals of the contemporary media environment. This is so because everything we do with new media involves a practice of some sort of cultural production (whether conscious or not), whether we are reproducing the status quo, creating new culture, or engaged in activism. When we engage participatory media tools, we are products as much as producers of media. So rather than function as passive audiences (which is often reflected in traditional media education models), we are practitioners, mediators, and participants whose role in the political economy of new media is negotiated and ambiguous. But this was also true before the rise of the internet. For example, in a study of teens and their personal media use, Steele and Brown (1995) formulated the media practice model, which argues that teens participate in selection, interaction, and application of the kinds of media they engage with. According to the author's research, the nature of practice is dialectical; media influence happens not in a vacuum but in the context of everyday practices. In this respect, media's impact is complex and influential in less formal ways.

Relatedly, Giddens' (1984) influential theory of *structuration* proposes that culture and institutions recursively influence each other, meaning that individual cultural practices significantly impact how institutions ultimately structure themselves, while at the same time institutions reinforce previous cultural patterns. Because of this dynamic, social change can be instigated on the micro level of individual practice (Reason & Bradbury, 2006). With the rise of social media, changes within a local setting of cultural practice have the potential for amplification through the media ecosystem (Christakis & Fowler, 2011; Leadbeater & Powell, 2009; Rheingold, 2002; Shirky, 2008). The spread of the Occupy movement and Arab Spring are examples of this phenomenon in action (Mason, 2012). Additionally, viral media such as *Kony 2012* (Invisible Children, 2012) and upstarts like WikiLeaks (wikileaks.org) are forms of nontraditional media that reverberate through the *networked fourth estate* (Benkler, 2011), demonstrating how the potential for activism, accountability, and social change can arise from participatory cultural practice. Thus, emerging media practices can be generative, creative, and encourage positive social change (Castells, 2012; Mason, 2012).

So how can sustainability be leveraged in such an environment? If commercial media is any kind of societal barometer, the pervasive use of nature tropes in corporate communications (such as in advertising and public relations), even if the goal is to greenwash environmental practices, is indexical (as in, if there's smoke, there must be fire) of a latent ecological zeitgeist. Adrian Parr (2009, p. 3) says it like this,

> As the public's enthusiasm for sustainable ways of life, environmental stewardship, and social equality grows, popular culture is rapidly becoming the predominant arena where the meaning and values of sustainability is contested, produced and exercised. To state the obvious, this is because sustainability is social practice. It is an instrument of knowledge formation; it is how a local context is narrated; it engages new and emerging social values and the energies driving these in dialogue with more traditional values and conventions, along with the habits and stereotypes underscoring these.

Even if the stated aim of many professional communicators is to co-opt green or sustainable culture, it is a sign that culture does indeed evolve. Rather than just shaping society, media dialectically respond to a global shift in values. Williams (1980) recognized this potential in his critique of hegemony theory when he argued that ideology is not monolithic. He problematized the view of economic cultural determinism by arguing that cultural activities involve residual and emergent social practices that are not always in lockstep with the prevailing economic modalities of a given time. For instance, many cultural practices, such as those that contribute to the cultural commons, do not directly correlate with economic production and can be tied to ancient human proclivities. This is an important perspective when we begin to look at emergent social media practices that contradict the logic of command and control industrial organizational structures that emerged during the Industrial Revolution.

Indeed, Heise (2008) argues that the internet is often used in popular culture as a synecdoche for planetary connectivity, an indication of an ongoing desire for community and connection that is out of sync with the prevailing dogma of individualism promoted by the mechanism discourse. For example, in the highest grossing film of all time, *Avatar* (Cameron & Landau, 2009), the alien moon Pandora is itself a kind of organic internet where its native inhabitants can connect with planetary consciousness. Not surprisingly, ecological themes are pervasive in sci-fi films and often challenge prevailing ideology. As Brereton (2005, p. 185) suggests,

> 'nature' and its co-present ecological sensibility can evoke a potentially subversive, even utopian, presence as opposed to the 'cultural logic' of contemporary Hollywood film.... In effect, the science fiction genre affirms that we are only truly human when we are in contact with what is not human.

*Avatar*, along with other zeitgeist popular culture such as *The Matrix* (Silver & Wachowski, 1999) series, repurpose sensibilities and mythological tropes that are far more ancient than the rise of modernity. This should serve as a reminder that aside from the negative impact of media, they also process social anxieties and create connective tissue for human communication and creativity. Not surprisingly, Hartley (2012) asserts that media's social function is now in the midst of a fundamental conceptual shift from a top-down ideological model to one in which media transform into creative industries enabling DIY/DIWO (do-it-yourself/do-it-with-others) modes of production:

> The emergent 'creative industries' are in the twenty-first century taking over the position that 'the media' held in the twentieth. The media were conceptualized as the 'enabling social technology' of *ideological control* for a mass society, but the creative sector may be regarded as the social technology of *distributed innovation*. (Hartley, 2012, p. 7, emphasis original)

As an advocate for green cultural citizenship, it is hard for me to argue against distributed innovation in media education, as long as innovation is contextualized by the need to maintain healthy living systems. Indeed, media ecosystems should be gauged by whether or not they are healthy or unhealthy, and by how this is recursively impacted by, or affecting, the larger living systems that media are part of.

## Media and the Cultural Commons

Media and living systems are two interrelated commons, a commons defined as, "all that we share" (Walljasper, 2010). As highlighted by Shiva's (1993) *monocultures of the mind* concept, our attitude towards the cultural commons (such as the shared knowledge we have about the world) and environmental commons (such as commonly shared water and air) can be discussed as interconnected. Shiva asserts that cultural beliefs and environmental perspectives are indistinguishable and are the result of a *cognitive space* acting upon the world. For example, single-crop farming and corporate intellectual property control stem from the same beliefs that economy and living systems should be subjected to unregulated markets and privatization. The monocultural mind is a mentality that "locks us into a mechanical worldview, based on mechanistic science, mechanical science, mechanical production, and a mechanistic economics whose myth of perpetual growth leads us to death, decay, and disintegration" (Shiva, 2008, p. 142). One of Shiva's key insights is that monocultural thinking is *parochial* and not epistemologically universal: "Dominant scientific knowledge...breeds a monoculture of the mind by making space for local alternatives disappear, very much like monocultures of introduced plant varieties leading to the displacement and destruction of local diversity" (Shiva, 1993, p. 13).

Agriculture as a monocultural cognitive space can be illustrated by the story of tobacco, a topic that was for several years at the center of the media literacy movement in North America. The tobacco plant is a good example of a conversion from a local resource respected for its medicinal properties into a disembedded export commodity demonized for its transformation into a toxic and addictive product. A plant native to the Americas, it was largely used for ceremonies and healing by First Nations people. European colonists converted tobacco into an export crop, so important to the foundation of the United States that the tobacco leaf adorns the one-dollar bill. A telling scene in Terrence Malick's (2005) film, *The New World*, a re-visioning of the old Pocahontas/Captain Smith tale, brilliantly illustrates this change. We see the conversion of Pocahontas into "Rebecca" through cultural and agricultural domestication. When the colonists' tobacco crops fail, she helps the domesticated plants survive by using her ecological intelligence to compost the soil (she adds dead fish); meanwhile, the tragic character of "Rebecca" in her corseted and constricted Western dress is also being domesticated in order to be presented at the king of England's court. As she becomes "civilized," her family's society is destroyed.

Three hundred years later monoculture tobacco is fertilized not with fish, but with petroleum-based fertilizer, which is destroying soil's biodiversity and contributing to contaminated runoff into the Gulf of Mexico now causing large dead zones in the ocean. Furthermore, commercial tobacco is converted into an addictive product that causes five million deaths a year worldwide (growing to ten million annually within the next ten years), with mortality climbing mostly in peripheral territories. Commercial tobacco is often treated with the same chemicals that can be found in Windex, rat poison, and gun powder. The thrust of tobacco's history is that, like most other globally traded commodities, there is a destructively systemic structure that plays out a typical pattern in which localities (where the product is actually grown and harvested) are disconnected from the international market. The multinational tobacco companies are themselves engaged in investment capital and are deeply embedded in, if not epitomizing, neoliberal assumptions of market fundamentalism (Harvey, 2005). This story fits the same model of water, food and any other kind of commoditized natural resource.

The media ecosystem can be conceived of as a kind of cultural commons in which there is a major struggle between the forces of enclosure and democratic potential. Bowers (2012) asserts that the cultural commons represent non-commodified cultural practices, such as traditional activities, skills, and knowledge embedded within communities. The cultural commons is how sustainable cultural practices can be shared and passed between generations. He is concerned about enclosure, which is "the threat from ideological, technoscientific developments, and the efforts of the market system to incorporate different aspects of the cultural commons into the market system—thus transforming what remains of community self-sufficiency into dependence upon the market economy" (p. 225). When greening media education, it is important to connect the biosphere, media ecosystem, and the commons, and to highlight how they are threatened by enclosure, because conventional market norms favor unregulated, privatized, and closed technological systems that endanger living systems (Evans, 2012).

Media and living systems can be compared and contrasted to probe their interconnectedness and how distinctive worldviews lead us to act upon these systems. For instance, when Rowe (2008, p. 139) connects Wikipedia with Balinese water temples, he argues that the impulse to work collectively for the common good is the same:

> Yet in reality the Web is just a new venue for the same human capacity that found expression in the water temples of Bali. It is a long way from one to the other, in time as well as space. But in both the rice fields and on the Web, social structures and social norms are doing jobs—creating and managing resources that are held in common—that conventional economic wisdom says only monetary incentives and private property rights can do.... both draw on a side of human nature that does not exist in the economics texts and that has fallen off the radar in western economic life.

Certain commons-oriented behaviors that are emerging on the web indicate that collective intelligence and sharing are a key part of human social life. Like dandelions in an abandoned parking lot, "It is as though something latent in human nature is breaking through the concrete of the corporate economy and the bureaucratic state" (Rowe, 2008, p. 139).

In addition, the issue of open and closed systems can be correlated with the analog of agricultural practice. For example, monoculture farming is part of a closed food system that favors corporate food monopolies (Pollan, 2006). It is governed by numerous control mechanisms and a form of science that does not allow for the open-ended integration of natural systems. Open systems are like permaculture gardens, which are organized in such a way as to work within the given conditions of local environments (Holmgren, 2002).

Permaculture is not controlled with pesticides, petroleum-based fertilizer, or laboratory-engineered seeds. It is open to the conditions of local ecology and interacts with unpredictable elements, such as weather, insects, and native plants. Since agriculture is humans acting with and upon the environment, the kind of agricultural approach one makes entails a worldview. Societies that engage in monoculture are far different in outlook than those that use permaculture.

The agricultural metaphors of monoculture and permaculture translate to a number of debates regarding the difference between open and closed media systems and gadgets. Zittrain (2008) compares the closed system of the iPhone ("iBrick") with Google's Android to demonstrate the difference between disempowered and empowered uses of technology. In the former, Apple forbids its users to fix or tinker with their gadgets and software. In the latter, Google allows its source code to be available publicly so that its systems can interact with as many as possible. Wu (2010) charts the rise and fall of media empires, examining how monopolies are essentially closed systems that shut down innovation and democratic uses of emerging communications technologies. Boyle (2008) believes there is an information ecology that is threatened by the enclosure of private enterprise. Lessig (2008) writes about the difference between *read-only* and *read/write* information economies, in which creativity and innovation are inhibited by narrow and closed systems of intellectual property law. In this respect, the interests of corporate media can be compared to the practice of monoculture, whereas vernacular and grassroots uses of media in alternative contexts such as Occupy Wall Street tend to be more like permaculture (López, 2012).

All of these approaches can be contextualized by a keystone essay, Barber's (2006) "Pangloss, Pandora or Jefferson?: Three Scenarios for the Future of Technology and Strong Democracy." In it he argues about potential scenarios facing the future of the internet. Pangloss is the status quo, and has a donothing stance in which the internet evolves according to the needs of governments, corporations, and consumers. If this scenario continues according to its internal momentum towards closed monopolies, we end up with the Pandora scenario. Pandora represents the digital divide, digital rights management, Great Firewall of China, increased repression and surveillance, and the use of anti-piracy measures to shut off dissent and peer-to-peer sharing. The Jeffersonian model represents hope and would entail the active participation of users to continue developing and utilizing the internet to open up democratic participation.

The status quo (Pangloss) has elements of both a cautionary and hopeful future, but we are at a crossroads and it remains to be seen how open or closed the future internet will be. In terms of the environment there are similar parallels. Closed monopolized media systems further the interests of those powers that refuse to solve the ecological crisis. Climate disruption has to be resolved through democratic processes and sustainable culture needs to be cultivated through sharing and connecting. The Jeffersonian scenario goes hand-in-hand with transitioning from the centralized and closed energy system of petroleum, natural gas, and nuclear power to the decentralized and democratic potential of clean energy (Rifkin, 2011). Of the three scenarios, green cultural citizenship is more likely to thrive in the Jeffersonian model.

## Environmental Communication and Green Media Studies

Environmental communication has its roots in rhetorical theory. It is a theory group with practitioners loosely connected to a variety of disciplines, including communication, green cultural studies, green studies, film studies, philosophy, and American studies, and schools of thought like phenomenology, ecofeminism, and ecolinguistics. An area of inquiry for environmental communication is how film, TV, news, and public discourse surrounding policy issues impact our understanding of climate disruption and other ecological problems. The central concern of environmental communication is to understand how human communication about the natural world impacts the environment and how groups make environmental claims in the public sphere. It also problematizes the use of terms such as *nature* and *environment* and how language constructs human relations with living systems. Additionally, as in critical theory, the field strives to intervene in order to promote cultural change and is normative in outlook. Environmental communication approaches are as varied as media studies and often overlap.

According to Milstein (2009, pp. 346–348), environmental communication explores material-symbolic discourse (such as debates around the constitutive causes and repercussions of the term *the environment*), the role of communication in mediating human-nature relations, and applied and activist theory. Environmental communication includes risk theory, ecojustice, rhetorical theory, cultural theory, and ecosophy, and in some cases, work by scholars dealing with ecologically oriented media approaches combine environmentally themed versions of the cultural or media studies model of analysis. For example, Anderson's (1997) *Media, Culture, and the Environment*, Cox's (2009) *Environmental Communication and the Public Sphere*, Neuzil and Kovarik's (1996)

*Mass Media and Environmental Conflict*, DeLuca's (1999) *Image Politics*, and Hansen's (2009) *Environment, Media and Communication* are primarily concerned with framing and rhetorical strategies of competing groups (citizen groups, NGOs, governments, and corporations). The analytical tactics they use are varied, but they often promote knowledge, information, and discourse awareness about environmental issues in the context of mass mediation, public policy, and corporate spin. DeLuca draws on rhetorical theory, social theory, and postmodernism; Anderson and Hansen draw on media studies (critical theory, effects, uses, and gratifications) and social constructionism; Cox is heavily weighted towards the public sphere; and Neuzil and Kovarik stick closely to standard framing and mass-media analysis.

Additionally, environmental communication is driven by ethics, with many works using environmental philosophies (such as deep ecology or ecofeminism) as starting points for their analysis. Along these lines, one of the increasing focuses of environmental communication is its orientation towards risk analysis, particularly around the dangers of climate disruption. Indeed, Coupe (2000a, p.5) suggests one thing green theorists have had to do is to retrieve a sense of the real that has been lost in postmodernist and poststructuralist discourses:

> The focus of any praxis is on the future; with green studies what is at stake is the future of the planet itself. Class, race and gender are important dimensions of both literary and cultural studies; but the survival of the biosphere must surely rank as even more important, since without it there are no issues worth addressing.

Cox (2009, pp. 20-29) asserts environmental communication is a *crisis discipline*, meaning that our current predicament calls for urgent action and normative ecological ethics. The issue of a crisis discipline becomes important in my final analysis of media literacy education discourses, because one of the key themes is that many media literacy practitioners are skeptical of educational approaches that have political agendas. However, these practitioners probably do not realize that they also have an implicit political agenda by reinforcing an unsustainable status quo.

**Three Ecologies**

Guattari's (2008) *The Three Ecologies* is an important link among ethics, media, and philosophy. Guattari suggests there are three categories of action and interpretation that fall under a kind of ethical intervention called *ecosophy*. These *three ecologies* are: *mental ecology*, *social ecology*, and *environmental ecology*. In terms of ecomedia literacy, I suggest that these can be divided into three broad categories: phenomenology (mental ecology), practice (socio-cultural ecology), and the material conditions of the world (environmental ecology). Guattari's assumption, along with Cox (2009), is that if we do not intervene in these areas to address the environmental damage from what he calls *integrated world capitalism* (IWC), then we as a species and civilization will likely not survive. His efforts to break out of media studies' original one-dimensional mechanism represent an inspired and important step towards holism.

**Green Media Studies and Social Theory**

In response to the general deficit of ecological discourse in social theory, Barry (J. Barry, 1999, p. 216) asserts that in order to green social theory there are four central assertions that need to be exposed, all of which are also applicable to media literacy education. They are

> a rejection of the separation of "humanity" and "environment"; a stress on the biological embodiedness and ecological embeddedness of humans; viewing social-environmental relations as not only important in human society, but also constitutive of human society; and a claim that social-environment relations are of moral concern.

The final point was already made: the state of our biosphere requires immediate action from all sectors of society, media educators in particular. But the first three arguments, though obvious within an ecocentric framework, are not intuitive to media literacy educators. Nonetheless, environmental communication addresses the manner in which culture determines what is *environmental* and what is not, offering concrete lessons for media literacy. For instance, though media literacy is well versed in the tactics by which media construct identity and ideology (such as issues around race, gender and sexuality), it typically eschews nature from its agenda.

In my view, the difficulty for greening media studies, and in turn media literacy, is the result of two fundamentally related biases in academia. The first is a general failure to problematize technology and the other is the disappearance of living systems in the general discourse of humanities and social sciences. To address this gap, media education can be greened by combining elements of environmental communication (Cox, 2009; Hansen, 2009; Milstein, 2009), ecological critical theory (Merchant, 2008), green cultural studies (Coupe, 2000b), green social theory (J. Barry, 1999; Jagtenberg & McKie, 1997), technoliteracy (Kahn, 2011), Guattari's three ecologies (Guattari, 2008), and the expanded use of ecological metaphors by social scientists to describe the complexity of relationships in media analysis

(Altheide, 1995; Liska & Cronkhite, 1995; Luhmann, 1989; Nardi & O'Day, 2000; Naughton, 2006).

## Bridging Media Studies With Green Media Education

Media studies is an important foundation for media literacy, but as Kendall and McDougall (2012, p. 1) put it, "Media Studies has obscured media literacy." What they mean by this is that previous media study perspectives have had too much influence on how media educators engage the contemporary media and education landscape. Indeed, my own background is grounded in media studies, which tends to influence my approach to media literacy. This often means emphasizing a critique of media's political economy. A negative legacy of media studies is to emphasize the mass-media model and to view learners as uniform audiences that are impacted the same, regardless of culture or geography. In terms of green cultural citizenship, the positive side of media studies is that it does encourage an ideological critique of media institutions and technology. Critically engaging the economic foundations of media is important for gaining insights into unsustainable cultural practices.

To transition beyond media studies, the following list summarizes media theories that can contribute to ecomedia literacy:

- Critical theory: Analysis of institutions, political economy, and culture industry helps explain the ideological orientation of corporate media
- Cultural studies: Circuit of culture model describing how institutions and cultural practice impact each other recursively
- Media studies 2.0: Argument for how social media and participatory culture is breaking down the paradigm of monolithic mass media
- Media ecology: Critique and explanation for how media are also technological environments
- Environmental communication: Environmental communication's explanation for how the concept of *nature* is socially constructed
- Green cultural studies: Makes a theoretical turn by engaging our relationship with living systems
- The three ecologies: Looks at media from three perspectives: social, mental, and material ecology
- Ecology metaphors: Though often incomplete, when properly contextualized, ecology metaphors are an important transition to ecological thinking

Ecomedia literacy differs from traditional media studies by placing environmental issues at the center of its focus. It builds upon tactics already developed by other media disciplines (such as those listed above), but an ethical focus and ecocentric orientation are primary. Ecomedia literacy takes into account the welfare of animals, ecosystems, biosphere, and biocultural diversity, whereas critically oriented media literacy is mostly anthropocentric.

CHAPTER FOUR

# Mapping the Media Literacy Ecosystem

A primary assumption regarding media literacy advocates is that when learners become fluent in the techniques and tactics of media persuasion and production, it should lead to some kind of active and attentive engagement with media. But when it comes to promoting green cultural citizenship, this may not necessarily be the case. As Orr (1994) warns about standardized education's impact on environmental perception, literacy does not necessarily translate into ecological responsibility: technically oriented education can turn well-meaning people into planetary "vandals." As Orr asserts, "Now more than ever…we need people who think broadly and who understand systems, connections, patterns, and root causes…. This is an unlikely outcome of education conceived as the propagation of technical intelligence alone" (1994, p. 23). Indeed, media literacy practices can normalize the separation between media and the environment, and in some cases narrowly define technical proficiency as the sole criteria for literacy. Indeed, the manner in which "media" are defined ultimately impacts whether or not media literacy practices can be greened.

One of media studies' most important impacts on media literacy education has been how it approaches media as an "other." Like "literature" in literature studies, *The Media* is a socially constructed category (Bennett, Kendall, & McDougall, 2011). From this perspective media are an abstraction "out there" that has particular characteristics defined not by their objective reality but by the social practices of practitioners that view media through particular lenses. For instance, though books are a kind of media, they are typically eliminated. Other examples of excluded media include oral traditions (live music, spoken word, poetry), alternative media (zines, blogs, independent film, public access television, activist media), and art (photography, visual art, dance). Indeed, the kinds of media that are part of the cultural commons identified by Bowers (2012) are rarely referenced in media literacy documents. Typically "media" are thought of primarily in terms of commercial, mass, and electronic media.

Subsequently, *The Media* becomes a constructed idea based on an array of unspoken assumptions drawing on previous theoretical models, namely those that have emerged from media studies and cultural studies. Because of this constructed view of media, "institutional practices of teaching about popular culture must be understood as a technology for the naturalization of specific reading and writing practices, particular ways of making meaning and understanding the world which are far from neutral" (Bennett et al., 2011, p. 4). This partially explains the exclusion of an ecological perspective from media education, because sustainability and the environment have remained outside of the media disciplines. Also, it colors media as an institutional force outside of ordinary people, which invariably puts educators in a situation where only critique is possible. This partially explains why there is such a large division in media literacy between teaching *about* and *with* media. The emphasis on textual analysis can lead to reductive and reactionary responses to media, which is reflected in the protectionist stance. From this perspective, many who teach about and study media have little space for promoting alternative cultural practice other than those based on a negative posture. I personally saw this divide during professional conferences or gatherings where there was a very distinct split between those advocating media arts and creativity versus those who emphasize analysis and critique.

In the following discussion, I attempt to suss out these debates by examining major approaches to media education in the context of an ecocritical critique. I will then apply this framework to my own study of North American media literacy organizations and practitioners.

## Aims and Purposes

The general struggle between media literacy practitioners in North America is the tension between those who advocate technical skills versus those who believe in critical engagement. This divide is reflected by two major strands of media education: functionalist literacy and critical literacy. Pure functionalist approaches are apolitical and tend to focus on education *with* media by incorporating new information and communication technology (ICT) into standardized curriculum as a technical patch without incorporating a critical or self-reflexive component that analyzes its ideological dimension. This applied approach is often associated with *digital literacy*. In this regard, education with media should not be confused with the sector of education called *youth media*, which has mediamaking as its core component but tends to be community focused and stresses art, social justice, and voice (Fisherkeller,

2011). Rather, the key here is that functionalist media literacy is usually situated in formal education settings, such as schools. Those techniques that teach critical engagement tend to be *about* media and usually constitute the most common media-literacy practices in North America. This approach focuses on the study and analysis of media messages, with a big emphasis on deconstruction. When it comes to media analysis, there is also a division between skills-based approaches that are apolitical and activist approaches that promote social change. Not all media literacy is either *with* or *about*; it can also be a continuum in which students engage in both activities (this is normally my approach).

Tyner (1998) suggests that historically media literacy has a lot in common with literacy studies, she but acknowledges that on the surface the term *literacy* is often contested because it is too closely related to alphabetic texts (for example, the Spanish term for literacy, *alphabetismo*, is aligned with print media and comes across as awkward when applied to electronic media). For Gee and Hayes (2011, p. 22), literacy gets caught up in an assortment of different discourses:

> People have claimed that literacy leads to more humane and more modern societies and smarter people. In reality, literacy has quite different effects on different societies. It has no one set of predictable outcomes. It has certain "affordances," effects that arise if the context is right. Otherwise, literacy's effects, like those of other technologies such as television and computers, depend on the specific contexts in which different literacy practices occur.... The effects of literacy depend on what people actually do with it. In some cases, more literate and educated people are more politically quiescent and accepting of the status quo (because they tend to benefit from the status quo). In other cases, people have used their literacy skills to challenge the status quo and engage in political activism.

Literacy experts have a lot to offer media educators. For instance, what is true of traditional print-based literacy can also be true for *multimodal literacy* (different kinds of media such as audio, visual, and multimedia): they all have the intention "to support basic life skills and access to social capital" through the ability to decode (read) and encode (produce) texts (Gutiérrez-Martín & Tyner, 2012, p. 2). To this end, Gutiérrez-Martín and Tyner (2012) promote the idea that what we should aim for is not a *new* literacy as implied by the various literacies being proposed (*media literacy*, *multiliteracy*, *multimodal literacy*), but a new *dimension* of literacy that is multimodal and global. Such literacy would incorporate digital literacy, information literacy, and media literacy, and would fulfill Buckingham's (2007) concern that literacy is not just a metaphor for skill or competence but a social practice.

Depending on the aims and purposes of the literacy method, different social practices will be encouraged. For example, digital literacy (a kind of literacy that emphasizes technical skills over critical thinking) fits well into the scheme of standardized testing and privatization of education because institutions can offer it without encouraging the critical aspects implied by a broader definition of literacy that would include examining ICT's ideological context. Digital literacy as part of a neoliberal approach to education favors those skills necessary for market-oriented pedagogy. Consider the language from the following publicity announcement from Connect2Compete, a nonprofit attempting to close the digital divide: "Digital literacy is critical to America's economic future, and possessing these skills is now essential for accessing the jobs and education opportunities that will enable current and future generations to compete in the 21st century workforce" (Connect2Compete, 2013, para. 1). The organization partners with major telecoms, cable companies, software companies, and a few of the largest media companies in the world to promote an uncritical form of media engagement. In the crudest sense, digital literacy becomes a means for enabling knowledge workers to master software, efficiently multitask complex datasets, or develop information literacy skills, but it does not encourage the multidisciplinarity inherent in media studies or sustainability education. When conceived as a technical skill, and in particular as a technologically mediated competence, literacy can reinforce the disconnection between living systems and technology because the owners and controllers of these technologies benefit from this disconnected awareness (Maxwell & Miller, 2012).

In my opinion, when it comes to formulating education policy, enforcing the dichotomy between *with* and *about* media favors curricula based on technical skills pushed by market-based solutions that emphasize consumerism over cultural citizenship. Buckingham (2007, p. 147) critiques the market-based approach by arguing,

> If we want to use the internet or games or the other digital media to teach, we need to equip students to understand and to critique these media: we cannot regard them simply as neutral means of delivering 'information,' and we should not use them in a merely functional or instrumental way.

Consequently, Share (2009, p. 126) maintains, "A new epistemological framework for literacy education is now necessary because of the rapid growth of ICT, the expansion of free market global capitalism, and the escalating and vanishing linguistic and cultural diversity that is changing social environments at local as well as global levels."

In different contexts, media literacy can encourage social practices that support critical thinking and empowerment. For example, the work I have done in Native American communities was designed to support the technical and social needs of the students and their community (López, 2008). Combining basic literacy skills based on textual analysis and critical thinking with grassroots media production helped students to critique and appreciate a vast array of media—corporate to independent—while at the same time engaging in creative and positive mediamaking. The media production aspect involved teamwork, collaboration, creativity, and community participation. The intention is to support the social practices of a community's cultural commons. Thus, literacy is always within an educational environment that informs particular worldviews. Media literacy practitioners in formal settings such as private and public schools have to contend with the external pressures that those institutions impose upon practice; those practitioners working in informal environments such as communities are more accountable to the needs of those communities.

According to Buckingham (2007, p. 150), literacy in a broader sense includes the capacity for reflexivity based on "analysis, evaluation, and critical reflection" that involves acquiring *meta-language* ("describing the forms and structures of a particular mode of communication") and *critical framing*. As an example of a more complete model of media literacy, UNESCO's approach calls for "information and media literacy," promoting core competencies based on the 5Cs: comprehension, critical thinking, creativity, cross-cultural awareness, and citizenship (Wilson, Grizzle, Tuazon, Akyempong, & Cheung, 2011). These reflect core competencies for lifelong learning,

> ranging from access to information to the capability to analyze it, produce it and distribute in a variety of forms, including the uses of information and communication technologies as essential tools for information gathering, learning and communication.... Information processing and digital competencies are also associated with searching, retrieving, sorting, storing, recording, processing and analyzing information by strategically using a variety of strategies to verify the source and discourse of the communication for each form of media communication (textual, numeric, iconic, visual, graphic, sound, etc.). This process also includes decoding and recognizing patterns of communication that can be applied to different situations and contexts. It includes knowledge of the affordances for different types of information, their sources, their placement and the specialized vocabulary used for each media and distribution network. (Gutiérrez-Martín & Tyner, 2012, p. 3)

Gutiérrez-Martín and Tyner (2012) warn that a restrictive definition of competency can become tool based, promoting "'know-how' for information management." By contrast, they advocate a literacy that addresses "all aspects, objectives, content, contexts and implications related to the presence and importance of media in our society" (2012, p. 6). They concur with Masterman (1989) that, "all media are constructions that represent beliefs, values, and biases that subsequently influence their reception. As such, new media, such as the Internet, social networks, video games and so on could be seen as educational agents" (Gutiérrez-Martín & Tyner, 2012, p. 4). In this regard, they believe a critical media literacy should "address the major ideological and economic interests around ICTs and to support the critically [sic] analysis of the political economies of media business and the role of audiences as 'prosumers'" (2012, p. 4).

## Content Versus Context

A key debate centers around the difference between content-based *empowerment* approaches that scaffold skills (Hobbs, 2011; Scheibe & Rogow, 2012), versus a contextual approach that focuses on an ideological critique of media (Kellner & Share, 2007; Lewis & Jhally, 1998). This is reflected in how the organizations I studied fell into two camps: core and periphery. The core groups, which could alternately be called "educationalist," tend to be content oriented. Periphery groups, which are "interventionist," tend to examine the ideological context of media. Share (2009) argues that the content-focused approach promotes the literacy part of media literacy, whereas the context-centered approach is about media as an ideological environment. Decontextualized text analysis corresponds with Meyrowitz's (1998) discussion of the *content* approach to media literacy, which emphasizes studying themes and ideas such as violence, race, or sexism or the *grammar* approach of examining media texts such as film edits and camera angles. The context-oriented school is closely aligned with critical media literacy (Kellner & Share, 2007) and critical pedagogy (Giroux, 1994), which in turn are grounded in strands of critical theory from both media studies and cultural studies. Regardless of these main differences, all of these approaches emphasize the analysis of messages as their primary methodology.

Tyner (2010, p. 4) cautions that an overly critical framework can end up in "moral panics" about sex and violence in the media that lead to protectionist responses that "are at loggerheads with traditional literacy theories that envision literacy as a pathway to social capital, independent thinking, and pleasure." Protectionists seek to *inoculate* children from exposure to *harmful* media and generally disregard the audience's ability to make sense of their media environments. Protectionists can become overly negative in their response to media, failing to see or even respect alternative and participatory media practices. Protectionists also exhibit mechanistic tendencies, viewing media as a kind of ideological syringe that injects belief systems into passive audiences. Subsequently, Buckingham (1991) eschews mechanistic approaches to media analysis:

> If the media are indeed powerful, this power cannot be seen as something which producers impose on audiences: rather, it is a power which depends upon the active participation of audiences. The "power of the media" is thus not a *possession* of producers but an unstable and contradictory *relationship* between producers, texts and audiences. (p. 30, emphasis original)

Arts-based approaches emphasize the creative aspects of mediamaking and are traditionally associated with youth media, radio, video production, video-game design, and animation. An advantage of the mediamaking strategy is to teach medium-specific characteristics, including grammar (such as film or sound editing). As such, arts-based media education can be an entrée into the pedagogies associated with medium literacy, defined as "understanding how the nature of the medium shapes key aspects of the communication on both the micro-, single-situation level and on the macro-, societal level" (Meyrowitz, 1998, p. 103). An example of micro-level medium theory examines the difference in impact for interpersonal communication that can result from messages coming from a text message versus a phone call. An example of a literacy of macro-level environments is comparing the different impacts of traditional text-based literacy, online media, and oral forms of communication.

In an effort to balance the requirements of critical thinking, literacy, and mediamaking, Buckingham (2007) calls for a method that is not just about technology and information, but approaches media as *cultural forms*. This can help bridge the tension between a neutral literacy approach with the didacticism promoted by the critical media literacy method:

> Media education...is both a critical and a creative enterprise. It provides young people with the critical resources they need to interpret, to understand and (if necessary) to challenge the media that permeate their lives; and yet it also offers them the ability to produce their own media, to become active participants in media culture rather than mere consumers. It therefore involves the rigorous analysis of media texts, in terms of the visual and verbal 'languages' they employ and the representations of the world they make available; the study of the companies and institutions that produce media, and how they seek to reach their target audiences; and the creative production of media in a range of genres and formats. (Buckingham, 2007, p. 145)

Buckingham's method includes four categories of inquiry: representation, language, production, and audience. As discussed in Chapter 6, this method is closest to the holistic approach proposed by ecomedia literacy because it offers different lenses for exploring the multidimensionality of mediated experience, but it still does not challenge the recursive metaphors and models of Western epistemology that permeate media education, in particular the focus on visual media. For that we need to explore emerging pedagogies based on participatory media practices (see below).

## Personal Lessons

In 2002, I was trained in a content-based method of media literacy that is angled towards protectionism. Eventually, I became uncomfortable with this approach while working as a professional media educator in Native American communities from 2000 to 2006. The essential problem is that the protectionist approach is one dimensional. From a green cultural citizenship perspective, media literacy needs to be grounded in place, something that ecoliteracy advocates have long emphasized (Thomashow, 1995; 2003). Media are not necessarily experienced in disconnected contexts. As I learned through my various interactions with the media literacy movement, local cultural realities are often disregarded or viewed as irrelevant to the specific task of media deconstruction. This was particularly evident in Native American communities where I often had to recontextualize media to make sense of the context of their own community realities. For example, many protectionist media literacy advocates demonize tobacco, but in Native American communities it is necessary to distinguish between traditional and commercial uses of tobacco. Native American community members have told me that media literacy activists have alienated them out of a lack of respect for their tradition of using and offering tobacco for ceremonial purposes.

I also learned that more attention should be paid to the subjectivity of students and how different communication forms have particular media ecologies. Traditional media literacy rarely distinguishes between how differently the forms of TV, magazines, film, internet, or radio influence content. In fact, they generally treat content as independent of the form it comes in. But as scholars from the media ecology tradition have noted, communications can vary greatly depending on their medium (Meyrowitz, 1998). For example, reading printed texts involves different cognitive and sensory experiences than multimedia, and written communications convey messages differently than visual or oral media. The Native American students I worked with were primarily versed in the oral tradition; their learning context is much different than students who are raised in print environments. In practice, this means that print-oriented educational approaches tend to promote text-based literacy as a response to multimedia, whereas I found that arts-based and storytelling practices work better in oral cultural environments.

In addition, while working in Native American communities I experienced a cosmological difference between my background in media studies and how indigenous communities regard the world. In short, I discovered media studies is intrinsically anthropocentric, as compared with the ecocentricism that guides the educational priorities of the Native American communities where I worked. This difference led me to question the epistemological universality of media disciplines.

Ultimately, as a result of working extensively with Native American, Latino, and Afro-Caribbean communities, I discovered that contextualizing lessons with local knowledge is absolutely necessary whenever engaging larger issues related to media and messaging. There was no formula that I could "plug and play" for instantaneous results. In each case there was lots of intercultural negotiation, which only works within an open and flexible communications environment. As Maser (1999, p. 57) illustrates, embracing diversity will enrich the possibilities for sustainable cultural practices:

> Cultural evolution, like all evolution, thrives in a context rich in diversity and complexity, wherein myriad opportunities for interaction exist. Self-organizing innovations can emerge out of such a setting as people search for ways to live consciously and sustainably in every sense of the word. These innovations become 'attractors,' which draw us out of the chaotic soup into further experimentation with social/environmental sustainability.

In my encounters with media literacy I sensed a kind of framing that has a subtle cultural prerogative that potentially homogenizes a mode of thinking and engagement, such as emphasizing the visual sense or technological progress, which, in my view, is more often than not designed and promoted by privileged Euro-American "experts."

## Mapping Media Literacy Worldviews: Summary of Research Findings

Whether media literacy is used as a tobacco-use prevention tool or as a means to teach critical thinking about the news, each educator approaches and utilizes media literacy within the context of his or her figured world. Nonetheless, I also believe there is a shared figured world characterizing the media literacy ecosystem. In order to understand the common assumptions of the members of this system, I studied the documents of seven major North American media literacy organizations and interviewed nine key practitioners. First, I wanted to explore the state-of-the-art practices of media literacy organizations and key actors in the field during 2012-2013. Second, I wanted to investigate how root metaphors influence practices. Third, I wanted to understand the pathways that are available so that media literacy can contribute to creating an ecologically sustainable world. In the following sections, my research overview gives a clearer picture of how media literacy educators imagine the world, and what kinds of actors are included and excluded from that worldview.

The first step in this study was to situate discourses within a social context by determining who the key practitioners were (individuals and organizations) and plotting their positions in terms of what they consider to be acceptable practice. In order to do that, I performed a multi-site situational analysis (Clarke & Friese, 2010; Clarke, 2005) to map the major positions and worldviews of a vast array of practitioners. Multi-site situational analysis is a grounded theory approach developed by Clarke (2005) designed to map the situation of a research problem. In the case of my research, the situation comprised an ecocritical assessment of media literacy practices in the context of my desire to bridge media literacy with green cultural citizenship. Subsequently, situational analysis enabled me to chart the media literacy ecosystem according to the parameters I defined as a researcher.

Once I plotted my key sites for analysis, I gathered data in the form of archival website documents from seven major North American media literacy organizations, including National Association for Media Literacy Educators, Project Look Sharp, Center for Media Literacy, MediaSmarts, Media Literacy Project, Media Education Foundation, and Action Coalition for Media Education. During the research I identified two clear groupings that described these organizations: core and periphery. Core groups are mainly focused on media literacy in formal settings, such as K-12 schools (public and private), but also offer professional development for diverse sectors, including public health, academia, church groups, media business professionals, and activist organizations. Peripheral organizations tend to work in community settings and often engage *media justice* issues such as closing the digital divide, campaigning for government regulation, promoting activism, and supporting communicative empowerment. Organizations in the core group strive to link media literacy with mainstream education practices and tend to be apolitical in their approach; organizations in the peripheral group are activist-oriented with a clear political agenda.

I also identified nine key practitioners within the media literacy ecosystem in order to perform in-depth, semi-structured interviews. These practitioners serve as "keystone species," functioning as authors, international scholars, curriculum designers, program directors, media literacy activists, consultants, and website developers.

I analyzed the data using the methods of qualitative media analysis (Altheide, 1996; Altheide, Coyle, DeVriese, & Schneider, 2010) and critical discourse analysis that focused on the study of conceptual metaphors (Barker & Galasinski, 2001; Coupland & Jaworski, 2006a; Fairclough & Wodak, 1997; Gee, 2011a, 2011b; Lakoff & Johnson, 1980; Machin & Mayr, 2012; Morrow & Brown, 1994; Todd & Harrison, 2010; Wodak & Meyer, 2009a). My methods are described in greater detail below.

**Data Collection and Analysis**

Based on my preliminary literature review, I begin outlining the media literacy ecosystem's borders by drafting situational maps for ecomedia literacy and media literacy. A key process of multi-site situational analysis, this method involves creating three different kinds of conceptual maps:

1. situational maps lay out the major human, nonhuman, discursive, and other elements in the research situation of concern and provoke analysis of relations among them;
2. social worlds/arenas maps, which lay out collective actors, key nonhuman elements, and the arena(s) of commitment within which they are engaged in ongoing negotiations, mesolevel interpretations of the situation; and
3. positional maps, which lay out the major positions taken, and not taken, in the data vis-à-vis particular axes of variation and difference, concern, and controversy around issues found in the situation of inquiry. (Clarke & Friese, 2010, p. 366)

These maps helped establish the key actors (human and nonhuman), major discourses (schools of thought, methods, approaches), data sites (websites, professional associations, educational programs, social networks) and positions (educationalist versus interventionist). These situational maps served as an overview of the major elements that act upon the media literacy ecosystem, such as key trends that are in the background impacting how people think about media literacy education. These issues included the emerging ubiquity of mobile media gadgets, rise of social networks such as Facebook, corporate media consolidation, neoliberalism, education policy, climate disruption, economic crisis, evolving concepts of cultural citizenship, sustainability, social justice, technology debates, views on literacy, and so on. Though these topics are rather broad, they did have a bearing on the research situation, serving as a general environment in which policy, theory, and practice are negotiated. In addition, these maps became part of the data under analysis.

After completing situational maps, I refined the big picture analysis further with *social worlds/arenas* maps that focused on discourse themes and actors. These maps helped reveal the relationship between actors and discourses, such as where particular scholars are situated in relationship to journals, key documents, and professional organizations. The social worlds/arenas mapping tool was something along the lines of a family tree, although perhaps less linear in structure, delineating where boundaries are negotiated between social worlds in relationship to media literacy. Differing social worlds included academics, community activists, K-12 educators, and policymakers. Additionally, in the social worlds/arenas maps, members of the media literacy ecosystem's keystone species and their discourses came into focus. Here it was possible to trace relationships among leading organizations and how they connect within the broader field of practitioners.

The final map was *positional* (taken and not taken), focusing on controversies, methods, theoretical differences, practical differences, and so on. For example, as discussed there are major media literacy camps that are divided along the lines of analysis *about* media versus those that focus on technical or artistic skills by working *with* media. Then there are divisions between practitioners who promote critical literacy versus those who believe in constructivist models. Some approach media literacy as if media are a problem, while others believe that learners should decide on their own whether media are good or bad. External positions that influence the situation, such as those coming from media studies, cultural studies, or critical pedagogy were also charted. Whereas the social worlds/arenas map showed who belongs to what organization and where advocates for particular orientations were situated, the positional map brought into sharper focus the actual debates and positions of these various actors.

Mapping enabled me to identify which research sites to analyze. Research sites are defined as sources of discourse data that delineate *regimes of practices* (Clarke, 2005). Selecting and choosing these sites was done according to the method of theoretical sampling, which "means sampling to develop the properties of a theoretical category, not to sample for representation of a population" (Charmaz, 2010, p. 375). Charmaz (2006) asserts that in this approach "the researcher seeks people, events, or information to illuminate and define the boundaries and relevance of categories" (p. 189).

Media literacy is rarely offered in formal teacher training programs but is instead largely fulfilled by a limited number of programs offered by nonprofits, graduate schools, and undergraduate degree programs. In the spirit of informal learning that characterizes many new media practices, increasingly there are also self-guided online communities of practice and expert practitioners offering professional development for media literacy on the internet. Consequently, the web has become a central networking tool for practitioners where materials for analysis are freely available and accessible, and therefore are crucial for studying practice. As Fairclough (1999, p. 75) notes,

> expert knowledge/discourses come to us via texts of various sorts which mediate our social lives—books, magazines, radio and television programmes, and so forth.... Modernity can be seen as a process of 'time/space compression', the overcoming of spatial and temporal distance, and late modernity is marked by a twist in that process which is widely referred to as 'globalisation'. The vehicles for this spatio-temporally extended textual mediation are the new media—radio, television, and information technology.

Organizational websites are accessible to anyone with an internet connection, and by articulating particular positions they are communicating to an imagined audience that inhabits a figured world. By openly advocating particular positions and debates, organization websites can be viewed as the public face of media literacy. Through their websites these organizations demonstrate that they view themselves as part of the public sphere. Accordingly, web pages and downloadable documents (PDF, Microsoft Word) were primary sources of analysis because they represent how organizations interact with the media literacy ecosystem. While it is true that several organizations under analysis are private nonprofits that approach media literacy education as a business by locking many of their resources behind a pay wall, every site that I analyzed had sufficiently open explanations for their particular philosophy and clear definitions of how they approach media literacy. I did not analyze any documents that are inaccessible to anyone with an internet connection.

Stemming from my preliminary literature review and situational mapping, I initially identified four provisional data sites that I believed were key for analyzing the media literacy ecosystem: organization websites, journal websites, curricula, and practitioner interviews. After a preliminary review of these resources, I reduced the data sites to organizational websites and interviews. Though journal websites and curricula were useful for contextual information, they ended up being redundant as sources for discourse data.

**Selecting Organizations**

Media education encompasses a variety of practitioners, positions, and worldviews, and without an initial boundary, there could be no limit to the kinds of sites eligible for analysis. The first step to narrow my analysis was to focus on North American organizations dedicated to media literacy education (as opposed to just media literacy services). I did so because I am most familiar with this practitioner environment, and it was my experience within this information ecology that prompted my research in the first place. Nonetheless, my literature review showed that practitioners in Australia and the United Kingdom figured prominently in general debates, mainly because in those countries media literacy plays a central role in educational policy. Indeed, I found that oftentimes innovative ideas come from outside of North America, in particular those coming from the U.K. and Spain, with UNESCO representing more broadly international formulations of media education. Despite my study's regional focus, some of the crucial frameworks I used in my critique of North American practices came from outside the continent. Indeed, in a globalized media system, national boundaries are less clear. Ultimately, as the findings from my practitioner interviews indicated, what shapes and makes geographical regions particular has more to do with national educational policy.

In determining the sites for analysis, one exclusionary factor was whether or not the organization was simply a service provider. There are dozens of media education programs that provide educational and consulting services but are not contributing to media literacy debates nor are they producing unique materials. A way to determine this was to see if an organization was actually creating distinctive resources or curricula that added to the diversity of the media ecosystem. The most important control, however, was whether or not the key metaphors used on the organization's website went beyond the saturation point established by the initial coding categories that emerged from analyzing key organizations. If the organization was not adding anything new, then it was excluded from my final analysis. Interestingly, this excluded most organizations offering media literacy services, for the majority of them show remarkable consistency and conformity to the positions of the core organizations.

## Document Research

Once the conceptual maps enabled me to become conversant with data sources, I began familiarizing myself with the kinds of documents to analyze, which included home pages, "about" pages and any pages that featured definitions, statements of purpose, or declarations of principles. Next I selected teacher resources, which contained primary assumptions and methods for media literacy instruction. During the document selection process, I finalized my theoretical sampling strategy and returned to the situational maps to identify any additional document sites that needed analyzing. I captured these web pages and documents using the NVivo software platform. After porting them into the program, I organized, sorted, categorized, and coded all the documents using the method of qualitative media analysis, which is an ethnographic approach to document research.

## Interviews

Nardi and O'Day (2000) propose that information ecologies comprise *keystone species*, those individuals who are significant nodes within the broader system (such as librarians in the information ecology of libraries). Using the format of semi-structured interviews, I interviewed nine individuals who served as a *panel of knowledgeable informants* (Weiss, 1995, p. 17). They were drawn from a pool of identified key thinkers in the media literacy education ecosystem, chosen according to their influence as authors, leaders, educators, and trainers. From my personal experience as a practitioner, I am aware of the key players, organizations, and their debates. There are also a number of scholars who are unaffiliated with particular organizations but are important contributors to general discourses. I made a point to include interviews with non-affiliated practitioners to serve as an analytical control.

## Critical Discourse Analysis

Text-based documents and interviews were investigated using critical discourse analysis, with an emphasis on identifying the situated meanings of metaphors and what they revealed about the figured world of media literacy educators. In the field of ecocriticism, discourse analysis is closely aligned with the discipline of rhetoric and the study of literature, but I am more aligned with Garrard's (2004, p. 7) approach, which reads "culture as rhetoric, although not in the strict sense understood by rhetoricians, but as the production, reproduction and transformation of large-scale metaphors." For Gee (2011b, p. 73), discourse analysis becomes a tool of inquiry for understanding taken-for-granted worldviews. A technique for doing this is to look at the situated meaning of key terms as a *thinking device* for asking questions, such as:

- What are the situated meanings of key terms?
- What figured worlds are being assumed?
- What does the implied figured world exclude? (Gee, 2011b)

These questions were asked repeatedly as I appraised the coded documents.

Given that my approach to the discourses was critical, I enhanced my method of inquiry by incorporating critical discourse analysis (CDA). If discourses "comprise participants, values, ideas, settings, times and sequences of activities" (Machin & Mayr, 2012, p. 220), critique in this context is "essentially making visible the interconnectedness of things" (Wodak & Meyer, 2009b, p. 7). According to CDA,

> language both *shapes* and *is shaped by* society. CDA is not so much interested in language use itself, but in the linguistic character of social and cultural processes and structures. CDA assumes that power relations are discursive…power is transmitted and practiced through discourse. (Machin & Mayr, 2012, p. 4, emphasis original)

Not surprisingly, CDA is also closely associated with critical theory and cultural studies, the underlying theme being that CDA is concerned with the social construction of reality, power relations, and ideological structures (Coupland & Jaworski, 2006b). Norman Fairclough and Ruth Wodak (1997), key theorists in the CDA tradition, summarize the key assumptions:

> CDA sees discourse—language use in speech and writing—as a form of 'social practice'. Describing discourse as social practice implies a dialectical relationship between a particular discursive event and the situation(s), institution(s) and social structure(s), which frame it: The discursive event is shaped by them, but it also shapes them. That is, discourse is socially constitutive as well as socially conditioned—it constitutes situations, objects of knowledge, and the social identities of and relationships between people and groups of people. It is constitutive both in the sense that it helps to sustain and reproduce the social status quo, and in the sense that it contributes to transforming it. Since discourse is so socially consequential, it gives rise to important issues of power. Discursive practices may have major ideological effects—that is, they can help produce and reproduce unequal power relations between (for instance) social classes, women and men, and ethnic/cultural majorities and minorities through the ways in which they represent things and position people. (Fairclough & Wodak, 1997, p. 258)

CDA was crucial for understanding whether or not media literacy educators were reinforcing anthropocentric and mechanistic discourses.

The analysis of metaphors was my focus. Metaphor analysis evolved from cognitive linguistics, which views metaphor as linguistic and embodied (Z. Todd & Harrison, 2010). Lakoff and Johnson (1980) identify more than a dozen different classes of conceptual metaphors, many of which were relevant to this study, such as ontological and container metaphors. In addition, of particular importance was analyzing the use of *metonymy*, which is "the substitution of one thing for another with which it is closely associated" (Machin & Mayr, 2012, p. 171). A metonym is similar to what semioticians call a *synecdoche*, which is when a part stands in for the whole. An example of a synecdoche would be an image of a satellite dish to represent the media. In discourses metonyms are metaphors that have similar qualities or properties to the concept they refer to. So in the phrase, "we need to get more media literacy in the classroom," the classroom stands for formal education. In the phrase, "TV shapes our culture," TV comes to represent a much larger system, which could include corporations and consumerism. Given the complexity and systemic nature of media, metonyms are commonplace in media literacy discourses. Indeed, because media are so complex, for the sake of communication it is necessary to simplify with metonyms. For example, take the following phrases with metonyms and compare them with their potentially situated meanings:

- "We need to get more media literacy in the classroom": We need more critical-thinking and inquiry-based skills in formal education environments that are based on testing and Common Core Standards.
- "Media shape culture": The technologically mediated distribution system of ideas and symbols reinforces certain beliefs of the owners of media organizations and impact the way audiences perceive themselves and their role in society.

In day-to-day vernacular, detailed phrases like these would be a mouthful, so shortcuts are necessary to expedite communication. But as noted by Lakoff and Johnson (1980), metaphors include and exclude ways of perceiving a situation, and understanding comes from whether or not those who are communicating share similar frameworks. In this sense, when the terms *classroom* or *media* are used there are implicit and situated contexts that give those metaphors their meaning. For instance, the first phrase referring to classrooms is more likely (but not always) to be communicated within discussions of core group practitioners. The second phrase will likely be used by both core and periphery groups, but their implied meanings will be different. In the former media might be seen as neither good nor bad, but in the latter media might be assumed to be a bad influence.

Metonym analysis was also useful for delineating the *place* of media literacy. One of the conditions of an information ecology is that it is tied to locality. While many of the organizations profiled in this study have regional or local ties, as a whole, media literacy practitioners are dispersed. As such they participate in what Lévy (1998) calls a *virtual community*. Likewise, given the dispersed nature of practitioners, in the spirit of Anderson's (1983) concept of *imagined communities*, I presumed that practitioners have a collectively imagined figured world that constitutes practices that situate media literacy in particular environments. By exploring the situated meaning of metonyms such as *classroom*, *school*, *society*, and *world*, the goal was to understand where these places are situated within the social context of media literacy educators.

## Results: Mapping Positions and Boundaries

An important theme was how respondents often used spatial metaphors to describe different *spaces* and *realms* of practice, which corresponded in an interesting way with my choice of the ecosystem metaphor to describe the parameters of media literacy. While a biological ecosystem is an actual physical place, the practitioners I interviewed saw themselves engaged with a variety of different figurative places, using words such as *realm*, *field*, *place*, *sector*, *domain*, *area*, *space*, *sphere*, and *world* to describe different dimensions of practice. Some words to describe the boundary of these realms included *threshold*, *entryway*, *gateway*, *curtain*, and *door*. In some instances, realms of practices have their own particular language, which implies that a knowledge domain can be like a foreign country.

My use of the term *spatial* should not be confused with the orientation metaphors that Lakoff and Johnson (1980) write about, which refers to a class of metaphors based on spatial orientations such as up and down, or far and close. Rather, my designation is similar to the class Lakoff and Johnson identify as container or building metaphors, which imply an inside and outside, or something that is also visible from the outside. In this respect, my informants used an assortment of corollary metaphors, such as having a *standpoint*, being *grounded*, having *sides* or perceiving a space that one can *fall* into. In some cases a conceptual metaphor can have many referents, such silo used to describe closed academic disciplines. A silo can be a kind of space, structure, and container. A *pocket of information* is an example of a container metaphor.

The use of container metaphors to describe different realms of practice reinforces the concept of disciplinary boundaries. Historically, the boundary between what is media literacy and what is not media literacy has been negotiated by different disciplines through scholarly debates in academic journals, conversations at conferences, via written definitions by organizations, discussions on social networks and listservs, and through other practitioner discourses (written and verbal). Some of the disciplines that have contributed to this debate include media studies, cultural studies, literacy studies, film studies, rhetoric, education, youth development, public health policy, community activism, industry professionals, and psychology. Since the 1990s, groups have formed to refine and develop media literacy's boundaries. Several of my informants stated that defining media literacy as a field is still contentious, citing a number of issues, including whether it is a social movement or a field, and whether it is vocational, critical, or educational. One informant expressed the view that small groups cannot define what media literacy is. Despite these discussions, there are still people who call themselves media literacy educators who practice and produce documents defining the work they do.

Having established that the media literacy practitioners I interviewed perceive practice as taking place in different realms, I want to explore how media literacy's particular boundary takes shape. In the following discussion, I will describe the salient characteristics of the system to provide a general overview of the ecosystem's boundaries. First, I will detail more clearly the difference between the core and periphery orientations. Next, I will explain how these orientations affect the aims and purposes of media literacy. I will then elaborate on a variety of *external* constraints that influence what is *inside* the system and how all of this relates to the use of conceptual metaphor particular to this system.

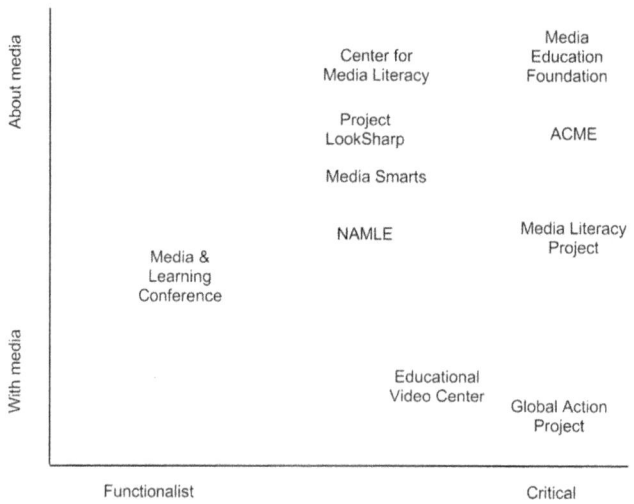

Figure 4. Orientations of media literacy organizations in this study. Their positions on the graph represent tendencies and not necessarily fixed positions.

## Orientations

As described, media literacy orientations tend to fall into four general positions (see Figure 4). Based on my multi-site analysis, I plotted these organizations onto a grid according to their public positions. The Y-axis represents a spectrum ranging from education *with* to *about* media. The X-axis represents a civic orientation with a continuum between functionalist and activist positions. It is likely that individual practices are more variable and that these broad categories gloss over subtleties. Indeed, my informants expressed a great deal more flexibility about practices that fall outside their domains than what is evident in the media literacy documents I analyzed, which often come across as rigid. Moreover, some of the practitioners I interviewed seem to embrace competing pedagogical positions. Nonetheless, overall this map helps determine where the majority of organizations are aligned. (Two youth media organizations, Educational Video Center and

Global Action Project, are on the chart because I originally planned to analyze their sites, but I decided to eliminate them from this study because their activities were too far outside of the practices of the main organizations being studied. Their positions on the chart indicate that their focus on production and critical action is different from the kind of work other organizations engage in.)

As a result of this research, I have categorized the media literacy ecosystem into two broad groups: core and periphery. The level of institutional support, functionalist orientation, membership numbers, focus on formal education, visible activity, and efforts to define the field demarcates the core organizations. Peripheral organizations are designated as such because of lower levels of institutional support, activist orientations, primary work in informal settings, and smaller influence in terms of defining the field. The core could alternately be called *educationalist* (based on skill development) and the periphery labeled *interventionist* (because of their intention to address social issues). Assigning the terms *core* and *periphery* was not a value judgment; neither was it meant to represent "better" or "preferred," but instead the terms were used to indicate level of influence. However, for the sake of transparency, I am more aligned with the activities of peripheral organizations, and the use of the term *periphery* likely reflects my own orientation within media literacy ecosystem.

## Core

The core group consists of MediaSmarts, Center for Media Literacy (CML), Project Look Sharp, and National Association for Media Literacy Educators (NAMLE). Not surprisingly, the founders of CML and Project Look Sharp were involved with the evolution of NAMLE and continue to remain active with the organization today. Through my discourse analysis and practitioner interviews, I found that the two common characteristics of core groups are that they emphasize integrating media literacy into formal education (in particular K-12 education) and stress skill-oriented and inquiry-based modes of analysis. Organizations at the core represent "mainstream" media literacy education because they are at the center of the struggle to define media literacy. Given that my informants for the most part believe that media literacy is a marginalized field, the term *mainstream* is somewhat ironic because of how media literacy has generally not been significantly incorporated into state education standards or teacher training programs. But I consider the core mainstream in the sense that the common position is more aligned with working within official education policy.

## Periphery

Action Coalition for Media Education (ACME), Media Literacy Project (MLP), and Media Education Foundation (MEF) represent the activist branch of media literacy practice, which places them outside the core of mainstream media literacy. Nonetheless, because members of these organizations are also significant contributors to key debates of the field, I believe they are well within the boundary of the media literacy ecosystem. Though there are tendencies within the periphery that are called protectionist, all of the practitioners I interviewed in this grouping were critical of and rejected protectionist media literacy, so they do not self-identify with the protectionist label.

Peripheral media literacy organizations fall outside of how core organizations define media literacy. For example, in its Core Principles document, the National Association for Media Literacy Education (2007) clearly describes practices they deem *not* media literacy education (MLE). Here are some key extracts (emphasis original):

> Simply using media in the classroom does not constitute MLE.... As a literacy, MLE may have political consequences, but it is not a political movement; it is an educational discipline.... While MLE may result in students wanting to change or reform media, MLE itself is not focused on changing *media*, but rather on changing *educational practice* and increasing students' knowledge and skills.... MLE builds skills that encourage healthy lifestyles and decision making; it is not about inoculating people against presumed or actual harmful media effects.... Making decisions for other people about media access or content is not MLE.

These exclusionary statements put several media literacy organizations outside the core group. For example, MLP, ACME, and MEF have all engaged in media activism, such as organizing against corporate consolidation of media and participating in national media reform conferences. Thus, because organizations in the periphery are engaged in media justice and have activist orientations, they fall outside of how the core groups define media literacy education practice.

But since peripheral groups do emphasize critically evaluating messages, I consider them to be inside the media literacy ecosystem. Indeed, the key insight I gained during my document investigation is that the analysis of messages is something that binds all these practitioners together. This is different than organizations with purely vocational or functionalist orientations that just teach information management or digital literacy, a type of education I excluded from my model of the media literacy ecosystem. This also differentiates them from youth media. Though youth media entails critical thinking, youth media practitioners emphasize voice and expression.

## Aims and Purposes

An emphasis on critical thinking unites the core and periphery groups, but it is the issue of aims and purposes that divides them. Ultimately, aims and purposes define what the practitioners mean by literacy. The following summarizes key insights from practitioner responses regarding the aims and purposes of media literacy:

- Those with a core orientation divide practitioners between those who are anti-media versus those with an education focus.
- Practitioners from both the core and periphery orientation rejected the protectionist stance.
- Despite the strong differentiation between anti-media activism and education, several expressed the importance of context and intent and how there is room for a diversity of approaches.
- Those coming from a peripheral and youth media perspective emphasized the importance of community involvement and a social justice framework, balancing voice and reading practices. Their aim is democratization of the media, and in one case a respondent believed a politically neutral stance is impossible to achieve.
- An important revelation from the interviews is how significantly the context determines the content in media literacy education. Some expressed that formal and accredited environments are going to exclude social justice and community-oriented approaches.

Coming from the perspective of aims and purposes, in general, the commentaries complemented each other. Each learning situation calls for an appropriate approach and tool. The rigid definitions and positions that have been posted publicly do not seem to reflect the private views of the practitioners I interviewed. Nonetheless, differences in approaches do exist, as evidenced by the discussion concerning a purely educational approach that claims to eschew agendas. All the informants claimed to believe that media literacy that is used for persuasion is not media literacy, and one suggested that literacy is a continuum. Finally, it appears that core practitioners view the periphery as protectionist, thereby conflating activism with a protectionist stance. Likewise, the periphery views the core as protectionist, perhaps seeing them as protecting educators from political involvement.

**External Pressures**

Media literacy organizations and practitioners are subject to external pressures that are not clearly visible in their public declarations on websites. Thus, the extent to which external factors outside of media literacy practices influence the field was one of the biggest surprises of my study. The media literacy ecosystem is embedded in a number of other ecosystems, such as education, economy, and technology, as well as the one that I care about the most, the biosphere. As such, these findings confirmed De Abreau's (2011, pp. 24-25) analysis that

> the research has clearly shown that the pressures of standardized testing faced by schools today have created some roadblocks. Indeed, most are feeling the enormous weight of carrying their current subjects and providing balance to their core curriculum. Added to this pressure are the economic tensions nationally and internationally which created severe cuts in the state and municipal funding resulting in the loss of school programs. The idea of teaching another form of literacy or its introduction into current curriculum causes many teachers to feel apprehensive, and furthering the divide is a lack of instructional training or tools to implement such change.

Additionally, in the practitioner interviews I was struck by the level of flux that is being felt directly and indirectly, pressures that would be undetectable in the written statements of media literacy organizations whose pedagogical strategies seem stagnant in comparison to the level of change that is taking place all around us. Indeed, as a kind of information ecology, many external technological, economic, and social pressures are disturbing the media literacy ecosystem:

- Participatory media breaking down the barriers between audience and producers
- Availability of smart phones with digital video, which makes anyone a potential mediamaker
- The presence of user-generated media sites such as YouTube that blur the line between media companies and mediamakers
- User-generated media undermining the protectionist stance because transgression is not only coming from media corporations in the form

of racist or sexist media, but from school kids uploading bullying videos or documenting their drug intakes
- The loss of funding for nonprofits
- The presence of new media corporations and software companies funding media education projects, challenging the old debates about corporate funding of media literacy where it was easier to define non-corporate spaces
- The rise of Common Core Standards and testing in K-12 education
- The policy push towards vocational- and jobs-related educational approaches that eschew critical thinking.

Frequently mentioned concerns include the difference between formal and informal education environments, the changing funding atmosphere (underfunded school districts, defunded programs, lack of capacity), the needs of K-12 teachers (time constraints, lack of teacher training, testing, Common Core Standards), confusion about media (difference between traditional and online media, different definitions of media, what do when student-generated media is deviant, emphasis on text-based content), disciplinary silos, and a disconnection between policymakers and administrators from the needs of students.

These pressures impact orientations differently, such as the difference between informal and formal settings. Formal settings are official sites where education takes place, such as K-12 schools, teacher training colleges, or graduate schools. Informal sites are community centers, arts centers, after-school programs, summer camps, etc. This distinction seemed to clearly define the difference between core organizations and those practicing youth media or promoting media justice. Those groups primarily working within formal education environments have to deal with the increased reliance on national Common Core Standards, which establishes what should be taught in schools. Moreover, the increased reliance on testing in formal settings means that media literacy educators have to develop assessment-ready rubrics that can be tested. A related issue is the disconnection that policymakers and administrators have regarding the nature of media literacy and why it is relevant in the classroom environment. Subsequently, when media literacy is not officially prioritized in schools, it is also rarely taught in teacher colleges because it lowers the demand for courses. Finally, teachers' time is constrained, so that is an additional pressure for including media literacy.

Another important external constraint is funding. For core groups, the funding of public education impacts the ability to conduct their work within formal settings. The funding climate hurts periphery group in a different way because the source of their revenue is not necessarily tied to working within the official system. Several informants claimed that while demand for physical spaces and services increased, there is decreased capacity. They say that funders are also questioning the need for media literacy and youth media centers when media is now so ubiquitous. In other words, funders wonder why youth need to be taught about media production when they're already making videos and posting on social networks. Most importantly, though, organizations are potentially impacted by the way funding influences particular orientations. Some informants believe those organizations that work in formal environments are likely to be better funded than those that are oriented toward social justice.

These external factors transform the media literacy ecosystem. For example, it is axiomatic that when an ecosystem is disturbed, the ecosystem transforms (Golley, 1998). As explained by media ecology founder Neil Postman,

> Technological change is not additive; it is ecological. I can explain this best by an analogy. What happens if we place a drop of red dye into a beaker of clear water? Do we have clear water plus a spot of red dye? Obviously not. We have a new coloration to every molecule of water. That is what I mean by ecological change. A new medium does not add something; it changes everything. In the year 1500, after the printing press was invented, you did not have old Europe plus the printing press. You had a different Europe. After television, America was not America plus television. Television gave a new coloration to every political campaign, to every home, to every school, to every church, to every industry, and so on. (Postman, 1998, para. 16)

In the following chapter, I will discuss how the discursive realm of the media literacy ecosystem does not correlate with current changes, and what can possibly be done about it.

Chapter Five

# The Media Literacy Ecosystem's Dominant Paradigm

My analysis started with the premise advanced by the New London School that pedagogy must "be based on views about how the human mind works in society and classrooms, as well as about the nature of teaching and learning" (Cazden, Cope, Fairclough, Gee et al., 1996, p. 82). This meant exploring for implicit assumptions about teaching, learning, cognition, and society. In some cases media literacy organizations do a good job of making their pedagogical theories explicit, but in most media literacy documents I analyzed the underlying assumptions behind their methods were only implied. Nonetheless, my research revealed that there are enough commonalities to generalize about the shared assumptions of learning and cognition in the media literacy ecosystem. To be clear, I am aware that most media literacy educators have individual differences in approach and worldview, which was confirmed by my interviews.

When I began my investigation, I assumed that primary actors in the media literacy ecosystem would be teachers and students, but as I conducted my research additional actors emerged and the implicated world in which they live expanded. The role of *stakeholders* is not always explicit in media literacy documents, but their presence is strong. These stakeholders include (but are not limited to) funders, policymakers, technology companies, media corporations, school systems, K-12 educators, parents, and youth. Moreover, the situated meaning of "media" is likely to imply corporate mass media and not necessarily alternative or community media. How media literacy educators view the function of media in society implies a civic orientation, which has a direct bearing on whether or not media literacy can be a tool for developing green cultural citizenship.

The coding categories that emerged during my research painted a picture of the world in which media literacy educators see themselves working. My rationale for these categories is as follows:

- *Media* is the most important concept for media literacy educators, yet poorly defined. My analysis sought to clearly define the shared (and divergent) meanings of the term.
- *Implicated actors* are the members of the media literacy ecosystem who are discursively created, such as students, mediamakers, policymakers, and citizens, but are not necessarily co-authors of the documents. These are the stakeholders of media literacy education.
- *Lifeworld* represents the implied subjective reality of the individual living in a mediated world.
- *Public sphere* is the category I used to delineate the realm of civic engagement. Here I wanted to explore situated meanings of commonly used terms like *citizenship* and *democracy*, which are often used without definition or clarification.
- *Literacy practices* have to do with those skills and aptitudes that a media-literate individual should possess.

In the following subsections I map these categories in more detail, demonstrating the range of possible associations that are used when practitioners communicate about media literacy.

## Media

The term *media* was the obvious starting place for analysis, not only because it is the key term that binds media literacy educators, but because Meyrowitz (1998) observed that there are widely divergent views of how to characterize media and the primary conceptual metaphor used (grammar, environment, or conveyor belt) serves as a kind of methodology for designing, researching, or thinking about media literacy. Below, I explore in more detail the dominant metaphors used for media, but here I describe the various terms used in media literacy documents that are related specifically to media, collated from both core and periphery organizations.

# The Media Literacy Ecosystem's Dominant Paradigm 115

Figure 5. NAMLE logo. By making the "M" the shape of a traditional television screen, the implication is that media are primarily visual and screen-based. Source: http://namle.net. Used with permission.

I looked for as many descriptive terms for media as possible in order to gauge the kinds of associations that practitioners have for media. Terms that describe media include literal elements (*message, subtext, topics, story*), techniques (*political rhetoric, spin, persuasion*), representations (*racism, stereotyping, violence*), opinions (*lies, misleading, pleasure, problem*), effects (*disruption, inform, inspire*) and meaning (*truth, bias, reality*). Subsequently I organized terms into four subcategories: content, grammar, types, and systems. Notably, the most frequently used terms for media formats refer to traditional, visual, and technological media. The middle and lower tiers comprise digital media such as *video games, ICT,* and *internet*. I regard terms such as *message, information, content, text, images, ads, news,* and *print* to be visually oriented. In contrast, there are limited references to audio, oral histories, sounds, pop music, radio, and music. If I am going to take the most frequently used terms as evidence of a dominant view of media, according to these documents the media ecosystem is predominately visually oriented. To reinforce this point, consider the name of one of the main core organizations, Project Look Sharp. In addition, NAMLE's logo has the M (which stands for "media") enclosed within the shape of a television screen (see Figure 5).

## Implicated Actors, Lifeworld, and the Public Sphere

The figured world of media literacy is more often implied than directly described. My goal was to piece together this world by identifying implicated actors, the lifeworld they inhabit, and the range of political engagement available to them. In some cases it was difficult to decide how to categorize something because it fit into all three categories, such as government and school. Implicated actors have lifeworlds and also a relationship with the public sphere. Based on the results described below, it appears that the primary focus and concern for media education is the experience and needs of youthful students, community, and citizens.

**Implicated Actors**

According to my ecosystem framework, the figured world of media literacy educators is composed of *members*. Membership is not just entitled to the most obvious actors, such as educators and students, but also for those implicated actors that influence the system in an external way, such as media producers (corporations, media business professionals) and policy advocates (FCC, citizen groups). Additional actors include institutions (media industries, funders, government, schools, churches), social groups (youths, teachers, parents, church-goers) and people in everyday life (individuals, people, audience, humans). As expected, *students* and *youth* occupy the top tier of most frequently coded terms. If I combine *teacher* and *educator*, this is the third most mentioned category (tied with *community*). There is a balanced mix between *citizens* and *consumers*, *adults* and *children*, and *audience* and *individuals*. If I combine, *corporations, business, owners, media companies, distributor,* and *media industries*, they rank as high as *people*.

**Lifeworld**

One of the most important implicated actors, aside from the media literacy educators, is the individual who is in need of media education, a hypothetical person out there in media literacy's figured world. How does this individual experience the world? This category is an effort to describe this person, but it is a little bit of a "grab bag" trying to fit together identity, behavior, experience, and habitat. Based on the terms I coded, the primary place of this individual is in the *school/classroom* where *belief, values, attitude,* and *culture* are paramount. Other important centers of activity are *work* and *home*.

**Public Sphere**

Initially, I categorized this group as civil society, but as I worked through the data it became apparent that terms in this category were more closely associated with the public sphere, which I define as the mediated space of civic engagement and citizenship. In this category, the most frequently coded terms are *citizenship*, *politics*, and *democracy*. I consider these to be fairly generic and without any implied politics per se. Terms that were specific to political positions, such as *justice*, *power*, *activism*, and *reform* ranked considerably lower, and are mainly present because of the work of periphery groups.

Not surprisingly, core and periphery groups have different conceptions of *citizenship*. Core groups communicate about citizenship in the context of *critical engagement*, *active citizenry*, *potential contributors to public debates*, and learning about the First Amendment. The terms *equality*, *justice*, *rights*, and *responsibilities* were less frequent. The language of periphery groups associated with citizenship was more direct in terms of advocacy. Periphery organizations want to *challenge Big Media* and *corporate media* by supporting media reform, *understand* and *manage media consumption*, and recommend ways to join groups and participate.

**Literacy Practice**

Initially, I identified this category as *skills*. However, as I coded it became clearer that skills were a subset of *literacy practice*. Practice encompasses those activities and skills that media-literate individuals should be able to do or have. Notably, the terms most commonly associated with the generic definition of media literacy (*access, analyze, evaluate, communicate*) are among the most frequently coded. Skills and practices are dominated by reflective activities (*critical thinking, analyze, evaluate, understand, question, identify, inquiry, decode*); however, there are also many references to active engagement and production (*communicate, expression, create, creativity, active, mediamaking*). It is also worth noting that *with media* and *about media* were coded equally. With the exception of *collaborative*, individual skills are the most frequently coded terms.

## Discussion

Almost all of the media literacy organizations that were studied in this research were formed in the 1990s. Since then many external factors have changed the media literacy environment, including interactive media technologies (ubiquitous mobile media gadgets, social media, wikis, YouTube), education policy (changing standards and testing), state of the environment (climate disruption, biodiversity loss) and economic factors (the rise of neoliberalism and financial crisis). In the 1990s the world wide web was in its infancy, and yet to be invented were the iPhone, iPod, cheap personal computers, and platforms such as Facebook, Google, Twitter, and YouTube. The industrial model of mass media was the dominant form of media in that era (by *mass* I mean low-feedback, abstract, audience-directed media). Like me, several of the practitioners I interviewed started out by working in the traditional media business in some capacity during the 1980s and 1990s. Others entered into media literacy from academia, psychology, and education. Though our backgrounds vary, the period that we entered into media literacy was substantially different from the current environment in which media literacy is now being formulated and practiced. The 21st century's rapidly evolving media environment is characterized by continued disruption and change. Yet, media literacy education's basic model of media is primarily based on 19th-century meaning design.

## The Figured World: Medialandia

Regarding the rise of new media, educators periodically use the term *digital natives* and *digital immigrants* to indicate the gap between young and old generations. These terms imply that old and new media regimes are different places. Based on my analysis of the common metaphors used in media literacy documents, the media as a whole is conceived as a separate place to which none of us are native. As such, it is completely disembedded from the material conditions of living systems. In the following sketch, I call this discursively created place *Medialandia*. It is a composite of how the realm of media is depicted with commonly used metaphors and phrases (in italics).

*Medialandia*

Medialandia is not *real*. Though it is primarily occupied and controlled by *business interests*, it is also inhabited by *producers* that have individual *points of view*. Occasionally one can find *independent* mediamakers, but they are generally not visible. Medialandia is human built, constructed with *information* and *entertainment technologies* with no material reality or relationship with the physical world. Nonetheless, its *landscape* can be *navigated*. Sometimes messages are *hidden below the surface* but its *environment* can be *scanned* in order to *shift focus* and to learn new things.

To understand Medialandia, it must be *accessed*. Subsequently, its many *points of view* are embedded in *messages* we *receive* that are *transmitted* or *transferred* to us in order to be *reflected* upon. These messages are *constructed representations* of our *world* and they need to be *decoded*. We can get *pleasure* from these *transmissions*. Medialandia also has the power to *transmit* emotions through *messages*.

When *representations* of us enter Medialandia, they return altered, sometimes as *stereotypes*. Some believe we are in danger of internalizing Medialandia's *view* of ourselves. There is concern that different *social groups* can be influenced by how they *appear* after their *representations* return from Medialandia. Though some believe Medialandia is *fantasy* and we must be *protected* from it, most consider it better to learn about it and to *construct* new knowledge about it, because each one of us is an *autonomous* being with a specific *identity* that is *constructed*. Thus we have the ability to *construct* our *own meanings*. The primary way to understand Medialandia is to *reflect*, *critically think*, *analyze*, and *discuss* this place. Skilled practitioners can then *construct* and *encode* their own messages and *transmit* them back to Medialandia. Others believe that laws should be passed to change the behavior of Medialandia's rulers/owners.

**Implicit Assumption: Mechanism**

The above composite offers a general description of how media is imagined, offering some salient characteristics from the discursively constructed world of media literacy. To be fair, it glosses over variation and individual practice and is meant to stimulate discussion. Interestingly, the one element that tied all organizations I analyzed within the media literacy ecosystem was the emphasis on critically evaluating media messages. As discussed in Chapter 2, the predominance of the message metaphor has huge implications for how media literacy practitioners formulate their worldview and constitutes the primary meaning design of the media literacy ecosystem. When *messages* that contain *codes* are *transmitted* to *viewers*, this essentially reinforces the industrial-era communication model that approaches communication as a mechanistic process of transferring self-contained packets of information from one autonomous mind to another. Some examples of frequently found mechanistic metaphors from my research include:

- *Container/Building*: Message, construct, conduit, constructed, shape, cornucopia, surface, grasp, hidden, immersive, content, through, access, decode, decoding, deconstruction, encode
- *Transportation*: Transmission, transmit, transfer, receive, bombard, disseminate, convey

- *Visual*: Representation, presented, point of view, reflective, reflects, picture (of world), scan, shift focus, portrayals, positive/negative light, perspective, exposure, viewer, watch, view, objects for reflection
- *Place/Space*: Navigate, landscape, environment, travel, guide, place, roadside, locate
- *Media/Message as Thing*: Knowledge resource, take away different meanings, the content of media, acquire content knowledge
- *Mechanism (General)*: Programming, impression, autonomy

The use of *place/space* metaphors implies messages move through three-dimensional space, inferring that the totality of media is itself a self-contained realm occupying Cartesian space. Thus, the prevalence of container and transportation metaphors reinforces a mechanistic model of cognition and communication, situating media literacy's figured world within the historical discourse of mechanism. There also is a correlation between mechanism and stressing the visual sense. From a media ecology (media as sensory environments) perspective, an emphasis on the visual sense is unbalanced and largely the result of our text-based heritage (Abram, 1996; McGilchrist, 2009; McLuhan, 2002a; McLuhan & Powers, 1989; Ong, 1982; Shlain, 1998). Interestingly, in this sense, media literacy's standard practice is driven by media technology because a print bias underscores reflection, critical autonomy, abstract thought, and isolated critical thinking (Bowers, 2000, 2008, 2012). In all fairness to media literacy educators, Western culture is visual oriented (Sturken & Cartwright, 2009), so it makes sense that their practices mirror the dominant culture. As one of this study's interviewees, Helen, stated, "If you look at the history of media literacy education, I think that people are more readers than writers."

**Closed Knowledge System**

I consider media literacy's text-based modes of inquiry to be unintentionally ethnocentric. As Rasmussen (2000) asserts, in Western culture we over-emphasize literacy-based conceptual language and devalue other cultural forms of expression, what he calls *pattern languages* that are context dependent. This relates to how intercultural communication scholars differentiate between *low-context* and *high-context* communication (Samovar, Porter, McDaniel, & Roy, 2012). The difference between the two is how much the communicator depends on words or nonverbal cues for meaning. Western culture is primarily low context and depends on explicitly coded messages. Communicators that are high context communicate according to environments and situations and are often associated with traditional cultures. This also correlates with observations about the difference between Native American students who are *field dependent* and Euro-American students who are *field independent* learners (Swisher & Deyhle, 1992). Students who are field dependent learn better from the "outside" world, such as from other classmates, whereas those who are field independent tend to learn and solve problems independently. This is not to say one is better than other, but rather to stress that when one form of expression is stressed over the other, it ends up excluding and marginalizing other ways of knowing. As discussed previously, I believe non-Western modes of perception will adapt better to 21st-century thinking (such as systems and pattern recognition) and will therefore be more conducive for developing the cognitive skills necessary for sustainability.

The bias of the visual also creates some confusion about what is media. The most commonly referenced media were visual and text-oriented. As an educator working in multicultural environments, in my practice I find it necessary to highlight other kinds of media, such as oral expression so as to not exclude the cultural reality of the students with whom I was working. Textual and content analysis are very important, so I am not disregarding them. And to be fair, discussion and dialogue are encouraged by the majority of media literacy practitioners I studied, indicating that they believe in the power and strength of sharing knowledge and developing meaning in social settings. Nonetheless, there is a difference between how core and periphery groups regard expression. Core groups emphasize the importance of free speech and the First Amendment, whereas the periphery groups emphasize voice. This is an important distinction, because as African communication scholars and activists have stressed, there is a difference between the right of free speech and the right to be heard (Campaign, 2005; Windhoek + 10, 2001). Free speech in an environment dominated by corporate or state media monopolies does not guarantee the right to communicate equally in mutual exchange. Marginalized peoples (women, youth, economically disadvantaged) may have the legal right to speech, but often have no social space to communicate, thereby limiting a plurality of communication. Sustainability requires a diversity of voice and expression, so the more multicultural an educational framework can be, the better it is for 21st-century problem solving.

In terms of reproducing preexisting ideologies, the media literacy ecosystem is situated within mechanism. Given the ecocritical critique of Western knowledge, media literacy mirrors a very narrow and culturally specific model of the world that should not be considered universal. In contrast, sustainability educators recognize that ecoliteracy requires engaging *multiple intelligences* and *different frames of mind* (Goleman, Bennett, & Barlow, 2012; Lappé, 2011). I do not consider the media literacy ecosystem as entirely exclusionary, but I do feel it narrowly emphasizes particular cognitive models that stress critical thinking and individualism over systems thinking and shared meaning. My research shows that media literacy education generally reinforces preexisting notions of progress and individual autonomy, thereby marginalizing a generative space for green cultural citizenship. Moreover, because core media literacy organizations are trying to harmonize with formal education, they are more likely to reinforce the preexisting belief system of mechanism. As Sterling (2004, p. 10) asserts, in formal education environments

> Most learning...is functional or informational learning, which is oriented towards socialization and vocational goals that take no account of the challenge of sustainability.... This has been reinforced in Western educational systems by the introduction of a managerial or instrumental view of education—which has paralleled economic restructuring in recent years.... This modernist educational paradigm derives from a broad social and cultural paradigm, which is fundamentally mechanistic and reductionist.... There is a poor fit between this dominant paradigm and our experience of increasing complexity, interdependence, and systems breakdown in our lives and the world.

My belief is that green cultural citizenship is based on a systems approach with cognition as embedded within living systems and communication as ecological. I believe trying to combine media literacy based on mechanistic assumptions with a green cultural citizenship framework based on systems thinking will result in a double bind (trying to solve a problem with the same kind of thinking that created it).

## Media Literacy and Environment

As the data indicate, with the exception of the curricula created by Project Look Sharp (Sperry, 2011) and documentaries produced by the Media Education Foundation, the figured world of the media literacy ecosystem is mostly devoid of the term *sustainability*. *Ecology* was used as part of a subheading on ACME's website, but beyond that the term does not exist in any of the documents I analyzed. *Environment* was used in several instances, but as a metaphor for the *media environment*. Indeed, when I asked practitioners if there was a connection between media and environment, several responded first by talking about cognitive, symbolic, and technological environments. It took some prodding to discuss environment in the context of living systems. To be fair, many said they were not prepared for that topic and so were unable to make the leap to another domain of knowledge. Once we had mutual understanding, several themes emerged, which are discussed below.

**Practitioner Care and Concern for the Environment**

Despite the lack of evidence in document discourses, every single practitioner I interviewed expressed emotional responses about their concern for the environment, including *worry*, *distress*, and *terror* about the state of our planetary ecosystem. Most cited climate change as a key issue they care about. While several said that they do minimal activities for the environment (e.g., recycling, buying local or organic food), the majority did not anticipate pursuing the topic in media literacy, but would pursue it in other aspects of their professional activities. Three respondents expressed a desire to collaborate with me in the future to develop environment-related media literacy curricula.

**Barriers to Environmentally Themed Media Literacy**

The majority of practitioners interviewed felt that sustainability was an emerging issue and would likely be part of media literacy in the future, but few saw a place for it in current practice due to many barriers:

- Disciplinary silos isolate modes of inquiry
- Lack of knowledge and complexity of environmental issues
- Fear of teaching unknown topics or "wisely" not choosing to cover unfamiliar material
- Too many external pressures (time, formal, standards, testing, funding)
- Formal media literacy is mostly practiced in English education, which is not the normal place where environmental issues are handled
- Sustainability is primarily part of the science domain
- Sustainability education is "persuasion," which does not pass the media literacy "smell test"
- Media literacy is "sealed off from the rest of life"
- Normative ethics can be a barrier to standards integration

- "Wigitization" of education policy leads to mechanistic curriculum development
- Environmentalism is viewed by society as anti-progress and therefore is anti-business

Within the core group, it seems that there are so many constraints on current practices there might be little motivation to introduce new and challenging frameworks.

As evidenced by these comments, a sufficient bridge between media literacy education and ecomedia literacy is lacking because their approaches are epistemologically different. Methods like applying traditional media education tactics to environmental issues (e.g., content analysis of news coverage of climate science or policy) are potentially incomplete and insufficient when they reiterate Cartesian models of cognition and communication. As Sterling asserts (2004, p. 28), "making adjustments to the existing system" represents first-order change, whereas "changing the educational paradigm" represents second- and third-order change. Sustainability education, which seeks to shift from mechanism to ecology, functions on the level of second- and third-order change. The difference between these levels of change is discussed in more detail in Chapter 2, where I explore the *iceberg model* of systems thinking.

Other barriers are tied largely to external pressures and the situated discourse of mechanism. The core practitioners are struggling to get media literacy institutionalized in order to make it part of state and national standards and taught in teacher preparation colleges. It is difficult to argue against more media literacy in schools. Given the stress that media literacy organizations are under, I am now more sympathetic to the reaction of the organization president I wrote about in the Introduction. Why add one more thing to learn, train, and teach in a field already crowded with external demands? Subsequently, I believe it is more likely that community-based periphery groups will take up sustainability since environmental issues are more likely to emerge in community settings than in formal educational environments (for example, youth media projects often deal with environmental issues because they are real problems in their communities). However, it is important to acknowledge that it was a core organization, Project Look Sharp, which produced the most comprehensive media and sustainability curricula (Sperry, 2011). Traditional media literacy analysis has a role to play in ecomedia literacy, but it needs to be enhanced by broader educational practices that do not solely focus on media messages. In the

Conclusion I offer an expanded vision of how media can be a form of sustainability education.

## Recommendations

The danger of analyzing media literacy documents is that they are snapshots of a particular moment when the expression of those ideas made sense. The world is not frozen in time and is far more dynamic than the document analysis suggests (this is why interviews were so helpful). Nonetheless, the dominant metaphors used in media literacy education discourses are clearly situated in an industrial model of communication and cognition, thereby reinforcing an anthropocentric and mechanistic worldview. The emphasis on visual media underlines a Western cultural bias based on print literacy. Perceiving the realm of media as a separate place also reinforces the belief that media are disembedded from the physical environment. The question is whether the current media literacy ecosystem can adapt and remain resilient in the face of drastic social changes.

In the following sections I offer some suggestion for how it might be possible to adapt to the 21st century and to transform current practices in order to move toward green cultural citizenship.

### Reconceptualize Language

When looking at root metaphors, *root* is itself a metaphor, suggesting that language is like a living system in which words and discourses can grow. Like all living systems, that means language can evolve and change. When Nardi and O'Day (2000) propose rethinking technology through the concept of information ecologies they are trying to break the cycle of industrial discourses that reinforce the perception that technology is autonomous from humans. Likewise, we should use new metaphors to respond to our changing world and evolve our practices by repurposing ecological language. The media literacy ecosystem is one proposal; however, I predict that securing wide adoption of this phrase will be long and difficult.

### Media as a Place

The use of metaphors that regard media as a place (such as an *environment* and *landscape* that needs to be *navigated*) is noteworthy. Given the strategy of ecoliteracy practitioners to use place as a mode of inquiry, this appears to be a good leverage point for introducing ecomedia literacy. However, there is a difference between viewing place from mechanistic or ecosystem perspectives. As in my discussion of the mediapolis and green cultural citizenship, media can be approached as an ecosystem in which learners are viewed as inhabited members. This alters the traditional media literacy framework, which views learners as being audiences or users that exist outside the media system (as is the case of my composite of Medialandia). As members of the media ecosystem, learners have rights and responsibilities that go beyond the acquisition of technical skills, thereby bridging some of the differences that currently divide media educators (such as among educators in the core group and activists in the periphery group).

Ecosystem can be either a container or systems metaphor. By asking questions such as "What is the boundary of the media ecosystem?" learners could explore whether or not it is a place or a system, and how it overlaps. To this end, I suggest that media be thought of as a kind of augmented reality with affordances, rather than a place that exists elsewhere. This would shift the emphasis of media literacy from the analysis of messages to include a broader discussion of the impact of media on our sense of time, place, and space. Moreover, learners would be encouraged to connect the relationship between media and their particular local environment.

**Crossing the Bridge**

In the next chapter (Chapter 6), I offer my ideal media literacy model in the form of ecomedia literacy. In this section I supplement that discussion with some tools and themes that could serve as a bridge between media literacy and sustainability education. Below, I describe some key themes and offer some additional suggestions.

Tools and technique that can incorporate sustainability cultural practices:

- Mapmaking/environmental documentation (using media to document learners' landscape of experience)
- Disarticulation of corporate frames of nature through semiotics and media deconstruction
- Mediamaking/storytelling (harnessing media tools to become empowered communicators in the learner's home community)
- Autoethnographies (journaling or webcam monologues for self-reflection)
- Convergence media practices (collaborative mediamaking)

- Digital storytelling for narratives of connection

Curriculum themes that tie to sustainability issues:

- Consumption and sustainability
- Corporate representation of nature in ads
- Evaluating the ecological claims of products
- Deconstructing environmental news and government framing of environmental issues
- Car ads and the matrix of the oil economy
- Food—fast food, nutrition, soda and sugar ads, etc.
- Bottled water marketing and labeling
- Health and the environment
- Greenwashing
- Climate change claims-making by competing groups
- Environment and social justice
- Animals in the media (including wildlife films)
- Alternative media and artist responses to environmental issues

**Intercultural Communication**

A theme throughout this book has been the overemphasis of Western modes of perception (mechanism) as a hindrance to sustainability education. I believe that intercultural communication can contribute positively to a shift in pedagogy. For instance, as I have experienced with my own work in diverse cultural environments, cultures have different learning preferences, including diverse ways of perceiving and processing information, such as field dependent versus field independent, cooperation versus competition, trial and error, and tolerance versus intolerance for ambiguity (Samovar et al., 2012). In my book *Mediacology* (López, 2008), I argued that new digital media require a right-brained educational approach, which is more common in non-Western cultures. Could adapting our practices to non-Western learning styles encourage sustainability? Pink (2005) believes we are transitioning into the "Conceptual Age," which relies on right-brained thinking. This means moving from thinking that is sequential, text-based, and detail-oriented to thinking that is simultaneous, context-based, and synthesizes the big picture. The difference between the former and latter is similar to those differences between 19th- and 21st-century knowledge. I believe much inspiration for sustainability can be found in the learning styles of non-Western cultures, and

I encourage media literacy educators to examine their cultural assumptions more closely.

**Discourse Analysis and Semiotics**

Key to media literacy is the deconstruction of messages (Hobbs, 2011; McDougall & Potamitis, 2010; Scarratt & Davison, 2012; Scheibe & Rogow, 2012). Often semiotics is used for studying representation, in particular, racial, gender, and cultural stereotyping. Animals and living systems are also used and stereotyped in a variety of ways (Corbett, 2006). Why and for what purpose?

Media literacy has also pioneered techniques for analyzing the way media frame issues, both visually and textually. Since discourse analysis can be applied to news and propaganda, green media educators can use this tool to examine how a critical issue such as climate change is covered in the news, or how to detect greenwashing (Hansen, 2009). Claims makers—from BP to Greenpeace—vie for public attention. What strategies do they use, and what systems enable some voices and not others (Cox, 2009)?

**Authenticity and Resonance**

Media literacy practitioners have mastered deconstruction, drawing attention to nearly 30 different persuasion techniques used to manipulate and hook our attention (López, 2008). The primary technique, emotional transfer, is represented by how marketers (or propagandists for that matter) generate feelings in order to transfer those sensibilities to brands. But the various emotions generated by sex, fear, and humor are tied to more ancient needs related to our connection with living systems. Media literacy could point out that when advertisers are manipulating our emotions, they are trying to tap into deeper desires for authenticity and resonance that can be fulfilled by activities that don't require consumption and could even tie into our primary need to connect with humans and nature (Stibbe, 2009).

**Economics and Ideology**

The critical analysis of ideology is usually applied in the form of critical media literacy and aims to challenge the claims made by corporations and governments (Kellner & Share, 2007). In the age of Occupy Wall Street, much attention has been applied to the way in which economic values are propagated through media. To this extent, it is absolutely necessary to examine those discourses surrounding growth, progress, and consumption, and how they lead to debt on multiple levels: personal, social, and ecological (Gilding, 2012). To what extent are both economics and ecology ultimately two sides of the same coin?

An additional dimension can be explored: different media promote a range of environmental ideologies—beliefs about how we act upon the world—spanning from anthropocentric to ecocentric perspectives (Corbett, 2006). What implications do these different worldviews have for ecology? Moreover, given that most media literacy aspires to greater democratic participation, it would be good to examine the kind of democracy we believe in. Is it anthropocentric, or could we work towards Shiva's (2005) concept of Earth Democracy?

**The Cultural Commons**

Educators pushing for media justice can link the enclosure of the techno-communication system by telecoms and media corporations with the enclosure of the cultural commons. IP law, anti-piracy legislation, and corporate mergers all have the effect of limiting democratic participation and access to cultural resources. Highlighting the importance of open culture, reformed copyright laws, and a less restrictive approach to sharing can also be tied to how corporations are privatizing food and water resources.

**Intertextuality**

People should not only think about ecosystems, but think like ecosystems. This means examining our mental models and learning to think in terms of systems, relationships, and connectivity (Bateson, 2007; Blewitt, 2006; Lappé, 2011; Sterling, 2009). Our social networks do this naturally, but what about media texts? Traditional media literacy tends to focus on single texts (e.g., an alcohol ad), but what if we looked at texts as if they were nodes in the media ecosystem? The way the web makes all texts open works does that for us. Consider how *Kony 2012* (Invisible Children, 2012) became a dialogue between many different texts produced by a vast range of critics and supporters; or how a WikiLeaks document becomes linked to a web of ideas and practices; or how the film *Avatar* (Cameron & Landau, 2009) made linkages to various genres and tropes from other films, and then how fans and activists remixed and spread various memes from the film to support environmental activism.

## Gadgets

As mentioned, media education programs rarely critically engage the tools used to make media. We should celebrate the creative process and promote the empowerment of mediamaking, yet we should not ignore the fact that the gadgets we use have an increasingly negative impact on global ecology and social justice (Greenpeace International, 2010; Leonard, 2007; Maxwell & Miller, 2012).

## Phenomenology

Most media literacy looks outwardly to ask questions about what media do to us. Sometimes the question is changed to focus on what we do with media. But what about the manner in which media influence our cognition? How does media engagement impact our sense of space, place, and time? What are the *splaces* we are engaging? How might this experience of extending ourselves into media networks impact our *sense of planet* (Heise, 2008)? For this, it will be necessary to incorporate the perspectives of the media ecology tradition, which focuses on these problems (Ellul, 1964; Illich, 1973; Innis, 1999; Mander, 1991; McLuhan, 2002a; Mumford, 1967) and to apply an ecopsychology perspective (Conn, 1995; Glendinning, 1994; Kanner & Gomes, 1995; Roszak, 1995; Shepard, 1995).

## Alternative Cultural Practices

There is a tendency among many media educators to focus on the negative aspects of mass media. But we also need to support positive media practices (Buckingham, 2003; Gauntlett, 2011) that build trust (as opposed to distrust). After all, media are a necessary means for solving problems. While I fully endorse critical approaches, I also would like to warn against too much negativity that leads to learners feeling powerless and victimized. We need to pull people towards aspirational solutions (Cloud Institute for Sustainable Education, 2011). This is a slightly different take on problem-solving pedagogies that focus on how to fix problems. Rather, we should encourage learners to create solutions. The difference is subtle but important. What we are aiming for is supporting lifelong learning skills that build towards sustainable cultural practices that can envision a positive response to a very wicked problem (M. Bateson, 2007; Blewitt, 2006).

**Cross-disciplinary Collaboration**

I believe that media literacy practitioners could collaborate with and learn from organizations such as the Cloud Institute for Sustainable Education and the Center for Ecoliteracy, which have been able to synchronize education for sustainability with some Common Core Standards (Cloud Institute for Sustainable Education, 2011; Goleman et al., 2012). Standards developed by the Cloud Institute that cross over with ecomedia literacy include sustainable economics, healthy commons, natural laws, responsible local and global citizenship, and the dynamics of systems.

There needs to be cross-disciplinary dialogue and collaboration. It would be greatly beneficial for organizations in the core group to collaborate with education for sustainability organizations such as the Cloud Institute. The Cloud Institute has already developed curricula and strategies for harmonizing sustainability education with formal standards and assessment requirements. I imagine that an encounter between the Cloud Institute and an organization such as Project Look Sharp could begin the bridging process. Media literacy educators and sustainability educators could create a one-week intensive in which each group trains the other in their respective methodologies.

**Ethics Not Persuasion**

Media literacy educators argue that aims and purposes determine the methods (such as scaffolding skills versus didacticism). I agree that scaffolding and empowerment are essential. I also agree that didactic media literacy is no better than persuasion and does not work. However, I also feel that it is necessary to have an explicit and transparent moral framework that informs practice. Just as journalism has a normative purpose to be accountable to the public interest, media literacy educators should not be afraid to commit to an ethic of care for the environment; nor should a commitment to fighting climate disruption be seen as an agenda.

**Future Research**

This research can be complemented by a systematic study of metaphors from other disciplinary realms, including sustainability education, videogame literacy, and digital literacies, the lack of which is actually a limitation of this study's results. It is possible to parse some metaphors from the literature review, but a more in-depth study of discourses is needed.

I predict that as our ecological crisis deepens and cultural patterns shift towards greater concern for climate disruption, educators will want to explore new strategies for incorporating sustainability into their pedagogical frameworks. I also believe that knowledge generated from my research can inform future interventions that combine media education and sustainability-related topics, such as those examining food systems, climate change reporting, social constructions of nature, and human-animal relations. Ultimately, the media literacy ecosystem should become more diversified as a result of these conclusions.

CHAPTER SIX

# Ecomedia Literacy

Given my ecocritical critique of the media literacy ecosystem, is it possible to combine green cultural citizenship with media literacy? In this chapter, I propose that ecomedia literacy is a potential solution to the challenges raised by this book's analysis. I first discuss how ecoliteracy, education for sustainability, ecological design, and ecopedagogy can contribute to an ecomedia literacy framework. I then outline the curriculum principles of ecomedia literacy and close with a case study in which the framework was implemented in an undergraduate digital media culture course.

## Ecoliteracy, Education for Sustainability, Ecological Design, and Ecopedagogy

A meaning system that incorporates sustainability will encourage practices that promote healthy, vibrant living systems. As such, Stibbe and Luna (2009) propose that 21st-century skills for sustainability consist of ecological intelligence, systems thinking (gaining holistic perspective), appropriate technology, appropriate design and cultural literacy. According to Blewitt (2009), media education plays an important role because "sustainability literacy, however defined, requires sensitivity to virtual realism, to media ecology, and to those ongoing processes through which we shape and are shaped by increasingly ubiquitous technologies" (p. 1). Stibbe and Luna (2009, p. 10) broadly define ecoliteracy as "the skills, attitudes, competencies, dispositions and values that are necessary for surviving and thriving in the declining conditions of the world in ways which slow down that decline as much as possible." Capra's (2005) definition of ecoliteracy is, "to understand the principles of organization, common to all living systems, that ecosystems have evolved to sustain the web of life" (p. 230). Accordingly,

> This involves a pedagogy that puts the understanding of life at its very center; an experience of learning in the real world (growing food, exploring a watershed, restoring a wetland) that overcomes our alienation from nature and rekindles a sense

of place; and a curriculum that teaches our children the fundamental facts of life that one species' waste is another species' food; that matter cycles continually through the web of life; that the energy driving the ecological cycles flows from the sun; that diversity assures resilience; that life, from its beginning more than a billion years ago, did not take over the planet by combat but by networking. (Capra, 2005, p. 232)

One of the harder things to acknowledge about ecological approaches to education is that it is difficult to quantify sustainable behavior. Capra (2005, p. 20) warns,

Because living systems are nonlinear and rooted in patterns of relationships, understanding the principles of ecology requires a new way of seeing the world and of thinking—in terms of relationships, connectedness, and context—that goes against the grain of traditional Western science and education.

Despite these obstacles, though, Capra offers a model for ecoliteracy that can guide curriculum design, including *contextual* and *systemic* ways of thinking consisting of several perceptual shifts: "from the parts to the whole"; "from objects to relationships"; "from objective knowledge to contextual knowledge"; "from quantity to quality"; "from structure to process"; and "from contents to patterns" (2005, pp. 20-21). Indeed, the Center for Ecoliteracy (www.ecoliteracy.org) in Berkeley, California, which Capra co-founded, has developed a number of successful programs within formal education environments, as has the Cloud Institute (http://cloudinstitute.org) in New York City. Some of these strategies are discussed below.

**Education for Sustainability**

A concrete framework is education for sustainability (EfS). Inspired by the United Nations' education for sustainable development approach (UNESCO, 2012), the Cloud Institute's (2011) model for EfS focuses on how mental models influence our behaviors in the world. As reflected in the iceberg model of systems thinking, EfS promotes educational approaches that deal with root causes below the surface of everyday actions. Its curriculum design begins with an *essential question*, which is a query with no definitive answer but "should provide a compelling and relevant 'hook' into the students' own experience and knowledge base; [sic] stimulate inquiry, and maintain student interests" (Cloud Institute, 2011, p. 61). The essential question should be "open-ended and should relate to and/or uncover the essential understandings, processes and skills—to help students further journey into their own thinking and gain news [sic] ways to view the world in which we all live" (2011, p. 61). An essential question at the root of ecomedia literacy is, "What design elements does a healthy and sustainable media ecosystem have?"

EfS utilizes backwards design, meaning that the curriculum design starts with the essential question to establishe a framework, then moves backwards to build in its rationale, standards, indicators, outcome, and assessment strategy. The curriculum then scaffolds exercises, activities, and topics to lead learners towards answering that question. This approach is designed to incorporate Bloom's higher-order thinking skills, which include the ability to analyze, evaluate, and create (Cloud Institute, 2011, p. 61). EfS is *of* and *for* sustainability: it embeds sustainable cultural practice while promoting awareness of what sustainability is about.

To address the principles of EfS, ecomedia literacy should incorporate activities that focus on content elements, skills (abilities), and behaviors. An example of content includes understanding the lifecycle of a media gadget, including how it is produced and disposed of. A skill includes knowing how to access the information concerning how a gadget is made, or deconstructing the marketing of gadget makers. A behavior is related to practice: how does an informed green cultural citizen use the knowledge of the media gadget's lifecycle to make wise, sustainable choices about media?

**Ecological Design**

Another source of inspiration is the work of ecological designers, such as the Ecological Design Institute (EDI), who apply an integrated approach to endeavors as diverse as gardening, architecture, community planning, and education. EDI's five core concepts for ecological design are: "solutions grow from place"; "ecological accounting informs design"; "design with nature"; "everyone is a designer"; and "make nature visible" (Edwards, 2005, p. 102). These core principles can be incorporated into ecologically designed media education programs. For example, Buckingham (2007) calls for a media education approach that is not just about technology and information, but investigates media as *cultural forms*, requiring the examination of four categories: representation, language, production, and audience. As outlined below, EDI's design concepts can align with Buckingham's four categories of inquiry:

- *Production can grow from place.* This is possible in terms of the learner's orientation as an individual, connected being and embedded organism in the world. Any media that she studies or creates will come from *somewhere*, so it will be key to identify the various points of reference she is working within, including community, landscape, and lifeworld. This would counteract the "placelessness" of traditional media literacy approaches.

- As a consumer and producer, she can make an *ecological accounting* of her work, whether it is in terms of the lifecycle of electronic gadgets and nonrenewable resources, use of open or closed software and media platforms, or the general role her media usage plays in the scheme of sustainability.
- As a mediamaker, the learner is a *designer*; as an individual making sense of media, she is a *meaning designer*. As a practitioner and designer she is part of a network of users and creators, and as a consumer is a target of prevailing messages about consumption, which are of ecological consequence. In the scheme of a media ecosystem, the role of designer and audience are united by her role as a member.
- We absolutely must *make nature visible* and when it is *represented* to study closely those depictions. We can also make nature visible by conducting audits of our media gadgets during our ecological accounting and also by applying ecocritical analysis to media texts.

**Critical Ecopedagogy**

By fusing ecoliteracy with critical pedagogy (Gadotti, 2010; Grigorov & Matias Fleuri, 2012; Kahn, 2010), ecopedagogy is one of the very few frameworks that combines environmental education with critical technoliteracy, and therefore comes closest to the issues addressed by ecomedia literacy. Its key formulation is to "reorient" towards an "Earth pedagogy," or "Earth paradigm," in which Earth is viewed as a living organism (Gadotti, 2010). This perspective opposes hegemony and Eurocentricism (Grigorov & Matias Fleuri, 2012), thus corresponding with my own critique of Eurocentric epistemology and my advocacy for ecocentric green cultural citizenship. Salient practices include a pedagogy that is participatory, democratic, action oriented, and critical. One major technique includes using a scenario building method with "backcasting," which is similar to EfS's model of backwards design around an essential question. The idea is to solve problems through participatory practices, such as action research, in order to build alternative scenarios for the future. It also advocates experiential education that is relevant to everyday life.

**Technoliteracy**

Media exist because of the material reality of technology, yet it is extremely rare to find this addressed in media studies or media literacy. Maxwell and Miller (2012, p. 10) assert, "Before there can be a story to analyze, a message to decode, or a pattern to identify in collective or individual media use, there has to be a physical medium, a technical means of communication." Consequently there is a lack of awareness that

> the foundation of media studies is machinery that is created and operated through human work, drawing on resources supplied by Earth.... Despite this fact, media students and professors generally arrive at, inhabit, and depart universities with a focus on textuality, technology, and/or reception; they rarely address where texts and technologies physically come from or end up. (Maxwell & Miller, 2012, p. 10)

As a result,

> Media studies abstains from deep analysis of technology's materiality in part because the field remains in thrall to two largely distinct but eerily compatible discourses: First, a cult of humanism adores the cultural devolution afforded by consumer technologies that generate millions of texts and address viewers and users as empowered. Second, a cult of scientism adores the mathematicization of daily life afforded by the digital and its associated research surveillance of everyday life. (2012, p. 20)

This has led to an overemphasis on the study of symbols and ideas, and a de-emphasis on the study of media's materiality: "...we need to look at what happens to objects as much as brains" (2012, p. 20). This problem is closely related to the *American technological sublime*, which is the idea that since the Industrial Revolution in the United States people have substituted a connection with the *natural sublime* with the experience of technological awe (Nye, 1994). This has led to an uncritical embrace of all things technological.

The technological critique entails examining technology's role in society as a cultural form and the materiality of media. This refers to both the ecological impact of media practice (such as the specific impact of media technology on the biosphere), as well as its symbiosis with industrialism. As Hughes (2005, p. 170) argues,

> Environmentalists educate the public about endangered species, the decrease in biological diversity, and the loss of sustaining habitat. On the other hand, persons informing the public about technological change and the state of the ecotechnological environment, especially the human-built part of it, are few in number. To participate, the public needs to become technically and aesthetically literate, as well as being morally concerned about the role of humans as creators.

Regrettably, as Bowers (2000, p. 183) contends, one will rarely find "the cultural mediating characteristics of technology" studied systematically in universities:

> The double bind can be simply stated: the one place in society where it might be possible to learn about the cultural nature of technology, other than how to promote its further development, is unable to challenge the myth that equates technological development with social progress. Indeed, public schools and universities are chief promoters of this myth.

But it is important to not confuse a critique of technology as an anti-technology or anti-media stance. As I argued in my book, *Mediacology* (López, 2008), I believe media literacy is ultimately about recognizing a *sense of place* within the larger sphere of mediation. Sense of place is both crucial and paradoxical when we discuss the media component of ecomedia literacy. On the surface, media are the antithesis of place, not only in their utopian (*no place*) depiction of the world, but physically, as well. Where is the place of television or the space in which a phone conversation take place? This ambivalence leads to some hostility by ecologists towards media and technology. For example, Traina (1995) believes that media and technological culture cause the extinction of experience and that urban living is a kind of artificial reality (p. 20). However, Traina's advocacy of mapping as a tool for bioregional education reflects a major blind spot in the treatment of media and technology in environmental education. Maps are a kind of media that help us visualize the environment in different ways. Moreover, technologies such as GIS and GPS in combination with PCs can be instrumental for visualizing our environments from meta-perspectives. I find the anti-city, anti-technology bias found in some bioregional thinking a barrier against shifting those very areas that need the most change (and influence), and hence a good reason that we should combine media and environmental education.

Technoliteracy has to be integrative as not simply about technological tools, but about technologies as systems of social and cultural practice. As Kahn (2011, p. 12) notes,

> It cannot be stressed enough: the project of reconstructing technoliteracy must take different forms in different contexts. In almost every cultural and social situation, however, a literacy of critique should be enhanced so that citizens can name the technological and ecological system, describe and grasp the technological changes occurring as defining features of the new global order, and learn to experimentally engage in critical and oppositional practices in the interests of democratization, ecological sustainability, and progressive transformation.

One way to do this is to keep in mind an "indigenized education" perspective of technoliteracy that takes into account community needs, eloquently summarized by Deloria and Wildcat (2001, p. 32) as TC3 (technology, community, communication, and culture) and P3 (power-and-place-equal-personality): "Stated simply, *indigenous* means 'to be of a place'.... To indigenize an action or objects is the act of making something of a place." What this describes is very much in keeping with the concept of an information ecology, which is always information and technology being used in the context of a situated place with particular social practices.

To this end, sites of potential sustainable information ecologies are public libraries. Blewitt (forthcoming) has researched a number of libraries that combine sustainable architecture design with open information systems. When they integrate with neighborhoods, local businesses, and open green spaces, public libraries can serve as "living rooms of the city." Libraries using sustainable green architecture can act as a kind of "public pedagogy" encouraging technoliteracy, while repurposing *oikos* to support community information ecologies. As a buffer against unsustainable neoliberal ideology, they promote the idea of a cultural commons, the value of sharing, social connectivity, and learning, potentially serving as "people's universities."

Ultimately Kahn (2010, p. 77) sees technoliteracy, in combination with critical media literacy, as a key component of ecopedagogy, with the caveat that

> we must move from critique to design, beyond negative deconstruction to more positive construction of high-technology. But rather than following such modern logic of either/or, critical ecopedagogues should pursue the logic of both/and, perceiving design and critique, deconstruction and reconstruction, as collaborative and mutually supplementary rather than as antithetical choices.

He adds, "In other words, people should be helped to advance the multiple technoliteracies that will allow them to understand, critique, and transform the oppressive social and cultural conditions in which they live, as they become ecoliterate, ethical, and transformative subjects opposed to objects of technological domination and manipulation" (2010, p. 78).

## Media Literacy and Ecopsychology

Ecopsychologists argue that unsustainable cultural patterns repeat themselves due to a lack of consciousness for how our personal lives are interconnected with living systems (Conn, 1995; Kanner & Gomes, 1995). In this respect, despite my earlier critique of visual practices in media literacy, traditional media literacy techniques can actually help raise awareness about sustainable cultural issues through using textual analysis. As an example, what follows is an illustration of what I had in mind when I proposed to Jerry Mander that media literacy could be a method for raising awareness of globalization. In this case, I'm applying an ecopsychological framework.

As an object-to-think-with, I will briefly examine a Spanish-language Pepsi television ad that ran in 2003.[1] The commercial features Colombian-born pop star Shakira in a 30-second spot that is comprised of 27 edits (about one every second). Though the shots pass quickly, they are loaded with symbolism that draw upon myriad images resulting from the collision between Latin America folk culture and global consumer capitalism. One of the key aspects of TV is its existence as a "nonplace" that is more akin to a hybrid reality than a specific place (Foucault, 1998; Meyrowitz, 1985). The ad is holographic in the sense that as a part of the system that produced it, it reflects all the qualities of that system in the same way that a cell will embody the architecture of nature. They are both fractals. Consequently, the ad allows us to reflect upon what Conn (1995, p. 157) calls the self-world connection with the ecological self, which means that we can take any object from culture and extrapolate from it the vast ecological network of relations that produced it. As such, the Pepsi ad embodies the world system's production matrix of electrical power, processed sugar, water, plastic, gender identity, cultural transformation, globalization, media demographics, ritual, adolescence, and commercial entertainment.

There are four major themes in the ad worth noting: cultural identity, ecological worldviews, globalization, and adolescent ritual. First is the media spokesperson herself, Shakira, whose name means "Goddess of Light." In this instance she is a goddess of electricity, but even more. Originally a dark-haired Latina rock star, in the ad she has transformed herself into a blond vixen pop archetype like Madonna, Britney Spears, and Christina Aguilera. In the act of her transformation from "Latina" to "American," Shakira emerges from a red, white, and blue globe representing both Pepsi and global capitalism emanating from the United States. Secondly, there is a two-second-long shot in which we see the entire concert space containing a stage shaped like a crucifix, which is right below the hovering orb of a spherical Pepsi logo. The globe conjures a spaceship, but the concert space itself is a cool blue-drenched cathedral with high gothic pillars and round porthole windows that vaguely resemble stained glass normally seen in a church. Given that the ad is in Spanish and the

---

[1] As of the time of writing, the video can be viewed at http://youtu.be/dnLQnTOCF6w

audience is clearly Roman Catholic, the allusion is clear: the fictional rock concert is really a Catholic mass, a legacy of the Spanish conquest. But the ritual's fictional location is not rooted anywhere in particular. The hall's metallic-looking hall could be the hull of a spaceship or an aircraft carrier. In the very least, it exists as pure data, for the image could not exist without being processed with CGI (computer-generated imagery). Such nonplaces are common in the techno-industrial society, with its marketing system constantly alluding to our alien status vis-à-vis nature. In regard to strange nonplaces like that of the Pepsi church, Mander (1995, p. 319) states:

> Remaking authentic communities into packaged forms of themselves, re-creating environments in one place that actually belong somewhere else, creating theme parks and lifestyle-segregated communities, and space travel and colonization—all are symptomatic of the same modern malaise: a disconnection from a place on Earth that we can call Home. With the natural world our true home-removed from our lives, we have built on top of the pavement a new world, a new Eden, perhaps; a mental world of creative dream.

In this "creative dream" Shakira is a cultural domesticator, imbued with all the implications of the Church, placing the sacred inside an institution, or in this case, inside a plastic soda bottle, and displacing the sacred from nature. Unfortunately, as part of the hidden curriculum that marketing and TV represent in our world, the cultural values resulting from ads like this betray Snyder's (1995, p. 47) axiom that, "'She's cultured' shouldn't mean elite, but more like 'well-fertilized.'"

The ritual that occurs during the course of the ad is a Pepsified version of Mass in which Shakira, the Pepsi priestess of electrically powered globalization, shimmies with her body for the mass audience to imitate, a dance move that is similar to the wave of the Pepsi logo. In the end, in an act not unlike transubstantiation, she imbibes Pepsi along with the worshippers. Instead of wine turned into the blood of Christ, we have the blood of Earth—water—turned into the black water of cultural and economic imperialism. The message to the dark-haired masses of the global South is that if you want to dream yourself American, then you must consume the precious fluids of globalization so that you can magically be transformed into a Westernized commodity version of yourself.

In addition, this is also a concert, one of the few remaining venues for mass ritual in the techno-industrial society available to adolescents. LaChapelle (1995, p. 57) argues for a sense of place rooted in ritual experience: "Ritual ... allows us to bypass the limitations imposed by the structure of our language." Rock, punk, rave, and hip hop were modernity's efforts towards that end, but these ritual spaces have been subject to capitalist enclosure, as this particular concert demonstrates. The enclosure of our tribal inclinations for dance and release have the additional effect of freezing culture in a perpetual state of adolescence, or stunted "ontogenesis," something that Shepard (1998, p. 26) believes is at the root of our ecological malaise. In short, ontogenesis is the coming-of-age ritual that enables us to bond with our environment. Without these rituals we become disassociated and disconnected with living systems. Additionally, it is important to acknowledge that advertising is a ritualistic activity of capitalism, because marketing stimulates desire by invoking sexuality (Shakira in provocative clothing), connection (group ritual), love (the happiness we get from consuming Pepsi), fear (of not belonging to the tribe), and sensuality (the physiological pleasure that comes from media consumption). Stibbe (2009) identifies the substitution of these desires for authenticity as pseudo-satisfier, dissatisfaction-manufacturing, and convenience-constructing discourses. In Glendinning's (1995) terms, ads are products of "disassociation" caused by the collective trauma of being separated from nature.

Sewall (1995) advocates for a kind of mindfulness to address the visual field, which is often dominated by advertising. Pointing out that 50% of the cortex is for visual information, she suggests, "From a pragmatic perspective, this means that perceptual practice can ameliorate cultural conditioning and psychic numbing by reawakening our senses and intentionally honoring subjective experience" (1995, p. 203). Consequently, "Skillful perception is a devotional practice" (p. 203), which I liken to a kind of media mindfulness. Recommendations for ecological perception are: "(1) learning to attend, or to be mindful, within the visual domain; (2) learning to perceive relationships, context, and interfaces; (3) developing perceptual flexibility across spatial and temporal scales; (4) learning to perceive depth; and (5) the intentional use of imagination" (p. 204).

Arne Naess' (1995, p. 73) concept of deep ecology seeks to redress a lack of ecological balance through promoting cultural diversity, the antithesis of Pepsi's message: "Deep cultural diversity is an analogue on the human level to the biological richness and diversity of life forms. A high priority should be given to cultural anthropology in general education programs in industrial societies." I believe media literacy, when done with the proper pedagogical approach, is a kind of cultural anthropology. In particular, if it places dialogue at the center of its process, it can invoke the kind of questioning called for by Naess (1995, p. 75):

The decisive difference between a shallow and a deep ecology, in practice, concerns the willingness to question, and an appreciation of the importance of questioning every economic and political policy in public. This questioning is both 'deep' and public. It asks 'why' insistently and consistently, taking nothing for granted!

## Ecomedia Literacy Curriculum Proposal

In this section I bring together the various issues I have been discussing into a practical proposal that can serve as a model for ecomedia literacy. First, I will offer a theoretical overview and then present an outline of how these ideas can be put into practice as curriculum design. I close with a case study that demonstrates how these concepts work in a real-world situation.

The New London Group's (Cazden et al., 1996) pedagogical approach to multiliteracy, which they call *designing meaning*, has two dimensions: *how* we teach and *what* we teach. *How* it is done is through a cycle of processes that scaffolds situated practice, overt instruction, critical reframing, and transformed practice (not always in this order). According to their model, learning takes place in situations within learning communities where co-learners are expert mentors and the educator serves as a facilitator. Learners begin with the tacit knowledge of their particular lifeworlds, which includes social practices. The job of the facilitator is to introduce new concepts that lead to a critical reframing of the learner's knowledge resources. The degree of intervention (which can come from peers or through overt instruction) depends on the situated practices of the particular learning environment. Ideally, critical reframing leads to transformed practice.

According to this multiliteracies model, *what* the learner draws on are three resources of meaning: *Available Design*, *Designing*, and *The Redesigned*. Available Design represents the available resources at the learner's disposal. These resources can be tacit knowledge and the affordances of a particular learning situation. Designing is how these resources are made sense of when new concepts are introduced. The Redesigned happens when new connections and transformed learning takes place. The knowledge generated from this process then become new Available Designs. The facilitator scaffolds new concepts, while the learner spirals around emerging concepts.

Such an approach is the foundation for a dynamic model of media education proposed by Buckingham (2003). He suggests adapting the multiliteracies methodology in order to generate dialogue between the learner's tacit and explicit knowledge about media. Beginning with the tacit understandings and situated practices of the learner's vast experience with media (Available Design), activities are introduced in order to critically reframe that knowledge by encouraging learners to question what they know (Designing), with the resulting new awareness (The Redesigned) a form of media literacy. Techniques include various tactics that both familiarize and defamiliarize media, such as autoethnography (such as media diaries), translation between different media languages (writing about talking or talking about writing), group activities that include debriefing and mediamaking, decentering (seeing oneself through mediated eyes), and simulation and role play (creating a newsroom or setting up a video shoot).

Buckingham (2003) further elaborates that this approach is tied to Soviet psychologist Lev Vygotsky's concept of social learning: that knowledge is socially constructed through symbols and language. This includes scaffolding and dialogue as a means for helping students "reflect systematically on the processes of reading and writing, to understand and to analyze their own experience as readers and writers" (Buckingham, 2003, p. 141). Students are encouraged to develop a *metalanguage* so "they can describe and analyze the functions of (media) language; yet it also emphasizes the importance of teachers engaging with students' existing 'spontaneous' understandings" (2003, p. 142). To this end, reflection enables implicit knowledge about media to become explicit. The goal, then, is to spiral between vernacular and theoretical notions of media. "In classroom terms this means that learning activities are predominantly socially organized and a media classroom would be expected to be organized around reflective tasks, socially produced steps for development, further reflection and evaluation" (Domaille, 2012).

As applied to ecomedia literacy, *classroom* needs to be flexible: it can be a formal classroom, informal setting, or online course. Regardless, the concept is basically the same: new ideas about ecologically sustainable media (theoretical and semantic concepts) are introduced through critically reframing the learner's vernacular uses of media. Moreover, in an effort to embed ecological thinking into the curriculum design, whenever possible ecological metaphors can be used to translate concepts into the media ecosystem framework. For example, the turn of phrase from thinking about media as media ecosystems invites new possibilities for making connections between media systems and living systems.

## Ecomedia Wheel

Ecomedia literacy's curriculum design is based on a heuristic called the Ecomedia Wheel (see Figure 6). The Ecomedia Wheel is a method of inquiry I developed that enables learners to investigate media based on exploring a *boundary object*, which is an object with a commonly agreed upon identity but with different meanings according to its context (Gieryn, 1983). For example, most will agree that an iPhone is a mobile media device that allows users to access the internet and to make phone calls. However, it can also be used to trigger roadside bombs, upload videos of police brutality, access social networks, or share photos of family members. In the context of the user, it affords different possibilities. Students will be encouraged tell the story of the boundary object according to four lenses: worldview, environment, political economy, and culture. Worldview relates to how media impact our perception of place, time, and space. Environment constitutes the physiological effect of media on living systems. Political economy has to do with those forces, such as globalization, that drive the implementation and production of media, including gadgets and devices. Culture reflects the collectively shared symbolic realm that negotiates how we view media and living systems. By looking at media from the view of worldview, environmental reality, political economy, and culture, the intent is to guide learners towards an understanding of how media interact with us systemically.

These four perspectives combined with the boundary object make up the Ecomedia Wheel's heuristic. Visually it is a circle divided into four quadrants, similar to a Native American medicine wheel or Buddhist mandala. The choice for using these four perspectives is inspired by holistic methodologies that look at phenomena from different disciplinary lenses. These analytical frameworks include integral ecology (Esbjorn-Hargens & Zimmerman, 2009), cultural studies' circuit of culture (Du Gay, Hall, Janes, & Mackay, 1997), three ecologies (Guattari, 2008), the learning wheel (Nelson, 1998), media mandala (Tyner, 2011), permaculture design (Holmgren, 2002), and indigenous science (Cajete, 1994).

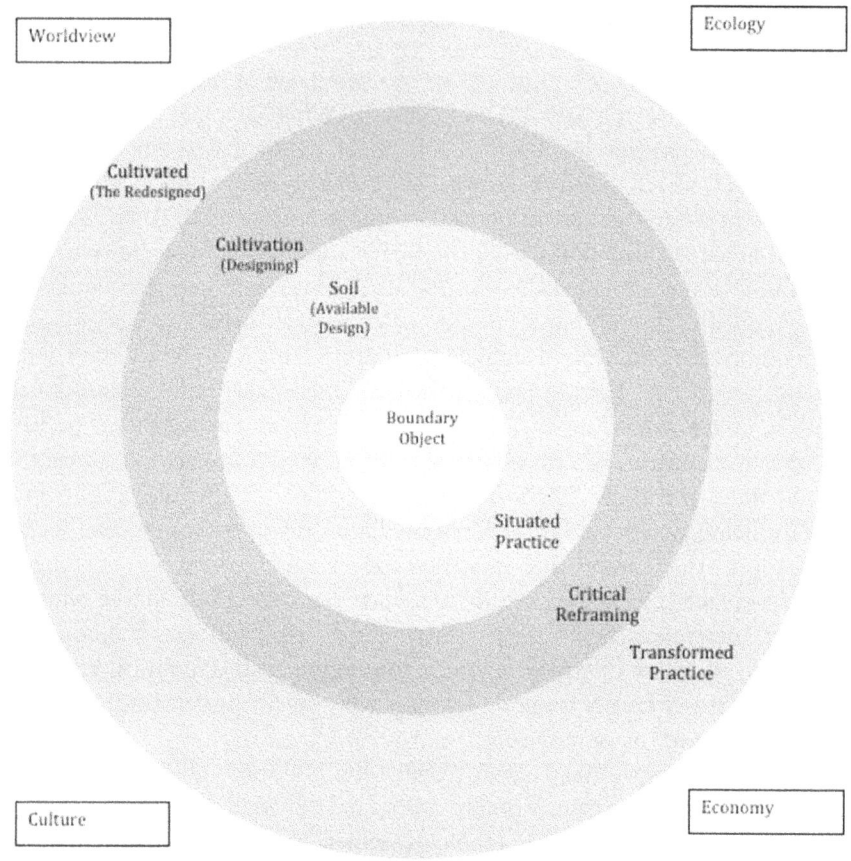

Figure 6. Ecomedia Wheel. The *boundary object* represents the unit of analysis (such as a media gadget like an iPhone). Each concentric circle represents a stage of analysis. The four corners marked by *worldview, environment, economy,* and *culture* represent different analytical perspectives.

## Soil (Available Design): Situated Learning

The integration of ecological metaphors works to support the critical reframing of tacit media knowledge about media. Pendleton-Jullian (2009) proposes that

> Given this interconnectedness of humanity and the natural world, theories, models, observations and experiments related to landscape and environmental ecology are proving increasingly useful to our understanding of other kinds of complex systems across diverse disciplines. This understanding allows us to think about change and resiliency dynamics, and it allows us to imagine constructing new models for change and resiliency. (p. 4)

Pendleton-Jullian suggests that the concept of ecotone can help us design appropriate educational spaces. An ecotone is the landscape ecology term for an edge environment, or zone between ecological systems, such as between a forest and meadow. This dynamic border region has aspects of its adjacent zones, but is itself unique. Changes in the ecotone's structure are caused by disturbances (a difference that makes a difference); in many cases such changes will impact the ecotone without significantly altering the core ecologies that border it. However, some disturbances can cause significant changes. In the case of education, a simple example of this could be a phone call received during a lecture. In this situation the learner represents an ecotone that borders personal life outside the classroom and the formal environment of the lecture. The call temporarily impacts the dynamics of the student's sense of space and may disrupt note taking and the lecture, but it doesn't alter significantly the individual's identity, the nature of the gadget, or the structure of the course. A major disturbance that would cause the student's learning ecotone to change significantly would be how the purchase of a personal media gadget draws the student's lifeworld into the classroom, including extended relationships, connectivity, and sense of space, place, and time.

Learners can explore a personal media ecotone through the individual's media gadget. Like the adage that a person who holds a hammer sees only nails, the person with an iPhone engages the world according to the possibilities the device affords. By augmenting our reality with a gadget, we do not move through the environment as disconnected subjects, as is the case in a mechanistic or Cartesian reality construct, but *through a world* of affordances that lead to shifting subjectivities. Wherever we are, the gadget will constantly shift our attention and change our relationship with the environment.

Media ecotones are similar to an information ecology in that they embody the situated practices of a particular information environment (such as a library or classroom). Nardi and O'Day (2000) believe that *information* does not represent disembodied facts, but is a mix of practices, social relationships, values, and technology usages within the specific reality of the practitioner. Nardi and O'Day argue that this nuanced understanding of technology enables an empowered stance in which media technology's influence is not autonomous or external from the user's interactions with it, as anti-technologists such as Ellul (1964), McLuhan (2002b), Postman (1993), and Mumford (1967, 1970) argue. For Nardi and O'Day, awareness of one's information ecology enables people to use "technology with heart," an important behavioral shift implicit in the call for sustainable cultural practice.

As it turns out, most of our education about media is through informal learning and social practices. Lifelong learning and media literacy are closely linked because our understanding of media is connected with how we informally acquire basic literacy (how to use and read media) just through engagement. The rise of social media and associated pedagogies destabilizes the normal conditions that constitute education in the formal sense, and highlight lifelong learning as an aspect of media practice. According to Blewitt (2010), this creates an opportunity for sustainability education to develop *ecotones* of lifelong learning:

> Networks of learning that can engage a multiplicity of participants, individuals, groups, organisations and sectors that may, or may not at first glance, share a great deal of common ground are nonetheless constitutive parts of a world of social learning enabling processes, spaces and practices of lifelong learning for sustainability to emerge. Thus instead of resembling sites of social and cultural reproduction, such a lifelong learning, and its constituent networks, must become an ecotone which is understood and lived both metaphorically and literally. To put it another way, a lifelong learning perceived and practiced as an ecotone is a transition area where different communities of practice, and interest, may come together thereby generating a richness in thought, action, knowledge, skills, understanding, creativity and philosophy not found within any one section, group, institution or community or in the wider educational environment. This transitional space offers the potentiality and possibility of rupture and a new ground for sustainability learning that is in essence politically democratic and just. It is the cultural space for a critical, border pedagogy. (p. 3470)

Learners' situated ecotones include the personal media gadgets and the social media they engage with on a daily basis. Students should be encouraged to explore these aspects of their lives because learner-centric education should not be about the world—as is the case in a traditional classroom where we study the *outside world* or media literacy that treats media as an external realm— but take place within it (D. Thomas & Brown, 2011). As such, the learning space should have built into it a pedagogical design principle of *situated learning* (Lave & Wenger, 1991), which is "situated within culture and context and cannot be separated out from this as formal education tries to do" (Weller, 2002, p. 75). Media, then, would always be approached within the ecological context of the learner's environment. The implications for choosing situational learning as a pedagogical strategy means embracing informal practice as a legitimate form of learning, shifting the teacher's role to facilitator, scaffolding skills through guidance, supporting community problem solving, and assessing for growth.

Advocates of situated learning believe that "learning must be personally meaningful, and that this has very little to do with the informational characteristics of a learning environment" (Mayes & de Freitas, 2007, p. 18). In the case of ecomedia literacy, co-learners are encouraged to make personal connections between their media practice and sustainability. This is made possible through the study of their personal media gadgets and how everyday media usage impacts living systems.

**Cultivating (Designing): Ecomediatone and Ecomedia Wheel**

The goal is to get learners to draw on their media ecotones as Soil (Available Design), which is then critically reframed (Cultivated) through analyzing their media gadget as a boundary object. As discussed, a boundary object can be assessed according to four perspectives: worldview, environment, economics, and culture. From the *worldview* perspective, a media gadget affords particular interactions built into the software design of the internet, social networks, information portals, and operating system of the device. These impact a sense of time, space, and place. Activities include understanding its impact on attention, including journaling usage, surveying the kinds of *feeds* the device is tethered to (social networks, "friends," organizations), and media fasting. These are all designed to highlight the tacit knowledge of the media user.

The *environment* perspective approaches the device according to its *lifecycle*—from resource extraction to disposal. The *economy* perspective situates the gadget in the global economy, both in terms of it being a commodity and as an enabler of ideology and social change. Additionally, the *economy of attention*—how our interactions are commodified (or not) is explored. Finally, the *culture* perspective relates to the cultural impact of the gadget, including how it affects status, language, relationships, empowerment, and the notion of citizenship and consumption.

These activities are embedded into a mediamaking task. This can be done with a free online presentation platform like Prezi, which is a multimedia tool with an "infinite" canvas that can embed hyperlinks, text, and multimedia (audio and video). Its nonlinear format allows users to diagram concepts and make connections in creative ways. For example, in order to learn the Ecomedia Wheel, Prezi can be used to embed links in a diagram that mirrors a circle divided into a grid of the four perspectives. Students can be prompted to answer a big-picture question, such as, "Using the Ecomedia Wheel, envision the characteristics of a healthy and sustainable media ecosystem."

**Cultivated (The Redesigned): An Ecomediatone for Global Responsibility**

The Ecomedia Wheel heuristic correlates with M. C. Bateson's (2007) model of global responsibility, which promotes, (1) systems metaphors, (2) crossovers between disciplines, (3) narratives of connection, and (4) crossovers with ways of knowing:

1. The Ecomedia Wheel's ecotone design is a systemic approach to media. The media ecotone, boundary object, and Ecomedia Wheel are systems metaphors for media. It translates tacit media knowledge into ecological metaphors.
2. With the built-in multiperspective approach of ecology, the Ecomedia Wheel enables learners to make connections across different disciplinary perspectives, including phenomenology, media ecology, critical theory, and cultural studies. This differentiates from standard media literacy practice because it incorporates a variety of lenses related to media rather than just focusing on textual analysis. Moreover, the pedagogical tools necessary to make these connections are multidisciplinary, including embodied cognition, social constructivism, situated practice, and connectivism.
3. The creation of a Prezi that integrates the various elements of the Ecomedia Wheel is a form of digital storytelling with a built-in narrative of connection. It tells the story of sustainable media through multimedia. Additionally, the learning environment is itself a narrative of connection through multimedia production and hypertext.
4. Participant observation and collaborative teamwork are integral aspects of the curriculum's design. This works well with the New London Group's (Cazden et al., 1996) multiliteracy strategy because it provides elements culled from the learners' personal lives (tacit knowledge), while at the same time introducing new concepts that then can be recombined during the critical reframing phase, with the final product reflecting new insights. As the New London Group suggests, the Cultivated (The Redesign) then becomes new material for the learner's Soil (Available Design). In other words, their knowledge gets composted!

## A Case Study: Greening a Digital Media Culture Course

In 2011 and 2013, I had an opportunity to put the ecomedia literacy framework into practice while teaching a digital-media culture course at an American liberal-arts college in Rome, Italy. In 2011, I taught two sections that combined for a total of 43 students, which I studied closely and present here, as a case study. Before 2011, I had taught this course three times, but this was the first instance I deliberately *greened* the course's structure and materials. My goal was to work within the university's standard requirements while introducing a green framework based on ecomedia literacy.

The primary method for this approach was the Ecomedia Wheel heuristic, which I used as a structuring device for the course content (a detailed explanation follows in the next section). I divided the 14-week course into six sections: introduction (including an overview of the Ecomedia Wheel), the Ecomedia Wheel's four areas of inquiry (worldview, ecology, political economy, and culture), and conclusion. Course activities included field assignments, online forum posts, Prezis (multimedia presentations), papers, and making YouTube videos. Online forums were located at the course's parent website hosted by me. We met twice a week in a physical classroom with each session lasting an hour and fifteen minutes. My method for evaluating the effectiveness of the course design was based on reviewing final papers, online forum discussions, anonymous student evaluations collected by the university, in-class interactions, and participant observation.

The catalog description (not written by me) detailing the Digital Media Culture (DMC) course is as follows:

> This course is an introductory overview exploring the ethical, aesthetic, political, social and economic dimensions of new digital media, which includes a critical discussion of Internet uses. We trace the history of digital media in the context of traditional media to understand their impact on society. We also seek to determine what the emergent properties of new digital media as they impact culture and society so that we can critically evaluate the various claims made about both the negative and positive social impact of new digital media.

Traditional digital media courses explore topics such as how digital media impact the media's political economy, audience participation, intellectual property, changes in perception of time and space, cultural citizenship, aesthetic practices, and social relationships. The main difference between my intervention and a traditional new-media class was the introduction of two themes. First, students explored how media gadgets impact their perception of the environments they inhabit. Students were asked to investigate the impact of devices on their sense of time, space, and place through a number of exercises, including media "fasting" and autoethnography using YouTube. Second, I had students contrast how digital media gadgets (such as iPhones) are represented in popular culture and marketing against the material reality of their life-cycle process (material production and disposal). They were also asked to apply all the concepts learned during the course to their personal media gadgets.

The primary textbooks were *The New Media Theory Reader* (Hassan & Thomas, 2006) and *Doing Cultural Studies* (Du Gay et al., 1997). Videos screened during class sessions were posted on the course website. What follows is a narrative description of readings and activities for each section of the course.

## Introduction (Weeks 1–2)

To begin the course, I wanted to introduce some guiding principles. For the readings, I assigned "Pangloss, Pandora or Jefferson?" (Barber, 2006), "Citizens" (Sunstein, 2006), the introduction to *Doing Cultural Studies* (Du Gay et al., 1997), and excerpts from *The Future of the Internet* (Zittrain, 2008) and *Remix Culture* (Lessig, 2008). The theme for the introductory sessions was to establish the idea of cultural citizenship and how it is important to understand the difference between open and closed systems. Barber and Zittrain lay out various scenarios that will result depending on how the internet and media gadgets are designed. The opening chapter from *Doing Cultural Studies* was used to introduce the circuit of culture concept and to supplement my explanation of the course's structure around the Ecomedia Wheel.

Students were also required to choose a personal media gadget to analyze throughout the course. I gave them the option of a smart phone, portable media gadget, personal computer, or gaming device. For the second week, they were required to keep a detailed gadget diary with the following instructions:

> Keep track of when, where, and why you used it, and note in what ways it was used in your life. At the end of the week, review your diary and then write a one- to two-paragraph observation of what you learned from your diary.

Students were required to post their responses to an online forum and submit to me a physical copy of their diary.

### Ecomedia Wheel Part 1: Worldview (Weeks 3–5)

This section began our exploration of worldview. Themes included technological determinism, time and space, and embodiedness. We started with technological determinism to explore the argument of whether or not media technology can shape perception. For the readings, I selected Carey's (2006) classic article about how the telegraph restructured American society; Mumford's (1967) deconstruction of the wristwatch; William's (1975) discussion of technological determinism and TV; and Cook's (2006) exploration of how print impacted culture.

Time and space were looked at through various lenses, including mobility culture (Green, 2006) and the consumer sublime (Nye, 2006). Embodiedness was explored through a critique of interactivity (Barry, 2006) and excerpts from *Digital Ground* (McCullough, 2004). *Digital Ground* examines digital technology from the perspective of interactive architecture and space —*spatial literacy*. As a corollary assignment, I asked students to "get lost" in Rome for several hours without the use of any gadgets and media, including pencils and maps. They were then required to post a reflection on the course website. The assignment was inspired by the radical arts group the Situationists, who made it a part of their creative process to get lost in urban cityscapes. Rome is a perfect place for this assignment because it has embedded within its environment many different kinds of spatial designs from throughout its 2,700-year history.

### Ecomedia Wheel Part 2: Environment (Weeks 7–8)

For this section of the course, students were asked to focus on the materiality of their gadgets. For background readings they read sections from *Greening Through IT: Information Technology for Environmental Sustainability* (Tomlinson, 2010), "Talking Rubbish" (Maxwell & Miller, 2009), and Greenpeace's online reporting mechanism for green technology (www.greenpeace.org/international/en/campaigns/toxics/electronics/). Students were asked to use the Greenpeace site as a starting point for looking into the lifecycle of their chosen gadget. In addition, I screened the documentary *Manufactured Landscapes* (Baichwal, 2006), which paints a vivid picture of the material conditions of manufacturing in China. I also screened *Story of Stuff* (Story of Stuff Project, 2009) to introduce an ecological systems perspective of consumerism. Additionally, we looked at various YouTube videos about conflict minerals, e-waste, and alternative design concepts such as cradle-to-cradle and biomimicry.

**Midterm**

For the midterm I assigned a short paper in which students had to evaluate the claims of gadget manufacturers by using the various concepts we examined during the first half of the course. I asked them to focus on Motorola's Xoom tablet by critically evaluating four videos produced by the company.

**Ecomedia Wheel Part 3 and 4 and Conclusion: Political Economy and Cultural Production (Weeks 9–14)**

The second half of the course focused on political economy and cultural production. Students were asked to read *Doing Cultural Studies: The Story of the Sony Walkman* (Du Gay et al., 1997) because it models the circuit-of-culture approach to analyzing gadgets. Though the book is somewhat dated, its method of research and analysis was intended to show students how to think about the complexity of gadget production, and to understand how culture, design, and political economy feed back on each other. To supplement and update the book's arguments for convergence media and convergence culture, I assigned chapters from Jenkins (2006) and Lessig (2008), and online essays by Kelly (2008) and Barlow (1993). I screened *Objectified* (Hustwit, 2009), a documentary about industrial design that features several interviews with Apple designers. We also tracked the evolution of the iPod's design through its marketing.

During this section students also began researching their gadgets for their final project, which had a written and multimedia component. The course concluded with the students producing a Prezi that tracked their gadget through the Ecomedia Wheel.

# Evaluation of the Effectiveness of the Course's Green Design

According to Wesch (2009), learning takes place when meaningful connections are made between semantic concepts and personal significance. For the purpose of evaluating this case study, semantic concepts are specifically related to the connection between media and sustainability. Personal significance means relating these concepts to one's personal media practices in daily life. There are two levels to this awareness. There is *local* awareness of how media gadgets impact time, space, and place on a daily level, and then there is the macro-level that represents broader, systems-wide understanding that connects personal usage with the global environment. I was also looking for evidence of basic media-literacy skills, such as the ability to critically read gadget marketing, find environmental information about gadget production, and to communicate these findings visually with Prezis and analytically in papers. These production skills (writing, making Prezis and YouTube videos) fall under the rubric of multiliteracy.

**Gadget Diary and Wander Assignment**

These assignments worked well together. The point of the gadget diary was for students to see patterns in their gadget usage. The wander assignment was intended to draw attention to patterns when their gadgets were not available. A few basic themes emerged from the diary assignment. First is that a majority of the students were not very introspective about their usage. They mostly reported it without much self-reflection. Common gadget uses, they wrote, were mainly for connecting with friends and family. Some students reported that they were surprised to see how addicted they were to their gadgets. Several of these students became aware of how gadget use is tied to *boredom* and *leisure*.

For the wander assignment, I contemplated the length of the assigned media fast. In past programs back in the United States, I had asked students to take a week off from media, but this was before social media and smart phones. Back then (ten years ago) it was a matter of turning off TV, and avoiding music and film. Now student lives are far more connected to the internet, especially in a study-abroad program. I realized that I could not make an unreasonable demand, especially considering how parents tether their children to communication devices. By assigning a one- to two-hour media fast, I think I was being a bit too gentle. But, according to the comments, even this short amount of time was difficult for many.

In general, there were two major reactions to the assignment: either students really loved it, or they hated it. Those who appreciated it revealed some interesting observations. Three mentioned that they smelled things for the first time. Others said they enjoyed listening to people speak Italian. Some marveled that Rome actually had interesting things to observe. Several commented on how they noticed that so many people walk around using devices. Many in this group said they would do it again (it remains to be seen if that is actually the case). A few students commented that they were glad they did the assignment early in the semester so that they could use the experience as a way of engaging the city throughout the term. The phrase many used regarding their regular gadget usage was *wasted time*.

The second group—those who detested the experience—wrote of unease, anxiety, fear, isolation, disconnection, and loss. These comments seem to confirm some of Turkle's (2011) conclusions about our increasing psychological dependency on gadgets. Not surprisingly, time seemed to really slow down for this group. Interestingly, those who hated the assignment were in the minority.

In general, I believe these assignments were very productive. In their final papers, many of the students drew upon these experiences to explain their deepened understanding of how gadgets impact their lives. I believe that these assignments in particular fulfilled Wesch's (2009) definition of learning, because whether they liked or disliked the assignments, they connected personal experience with the concepts of gadgets impacting time, space, and place from the course.

**Final Gadget Analysis**

An important component of the course design was to have students personalize their learning by applying course concepts to a personal device throughout the semester. At the beginning of the semester, students were required to track their personal media gadget in relationship to the major themes of the course's structure. Their final papers explored the gadget from the four dimensions of the Ecomedia Wheel. They were also required to apply these concepts to a multimedia presentation using the Prezi platform. The goal was to ensure that their learning combined multiliteracy and green cultural citizenship. Multiliteracy was evidenced in their ability to analyze media texts, use online communication tools, self-reflect with YouTube, and research the internet for information about how their gadgets were produced. Green cultural citizenship was demonstrated through their understanding of how their media gadgets impact the environment and their experiences of time, space, and place. Whether or not they act on this knowledge beyond the class remains to be seen. The ultimate test of green cultural citizenship would be if the students engaged their media technologies outside of class with their new awareness. Unfortunately, I was unable to track their activities beyond the class.

I was most struck by the YouTube videos. Students were asked to record themselves with a videocamera while answering two questions: "How does it feel to record yourself and to speak into a computer?" and "What is the impact of your gadget on your perception of time and space?" Their responses were remarkably candid and intimate, going far deeper than anything I have experienced while teaching an undergraduate course. Interestingly, the video recordings made students appear far more human than through the normal evaluation of written papers. I found their candid responses quite moving, actually. Many comments in their papers remarked about how difficult it was to do the video—technically and emotionally. Curiously, of all the assignments, this one was complained about the most.

In their written work, there was evidence of an awareness of the ecological dimension of their gadgets. However, upon reviewing their final papers, only a small number indicated "breakthrough" awareness about the ecological dimension of their devices. Many papers reflected understanding of the assignment's goals, but their writing did not generate any particularly enthusiastic calls for change or green cultural citizenship. Many wrote somewhat mechanically, reflecting the perceived desires of the professor. However, one key awareness that seemed to resonate with most students was that they understood the difference between open and closed media systems. Unfortunately, few related that with environmental themes.

Of 43 papers, only ten reported significant new awareness. These students connected green cultural citizenship with transparency and open systems. This was particularly common among users of BlackBerry, which has the worst environmental track record among gadget companies and is the most opaque in terms of reporting manufacturing processes. Many of these students commented on their previous lack of awareness and stated, had they known better, they would have made better decisions based on practicing green cultural citizenship.

**Course Structure**

Using the Ecomedia Wheel to structure the digital-media course helped students to more easily understand it because they had to "walk" through the entire process of the wheel (rather than just think about it abstractly). Admittedly, when I organized the course around the Ecomedia Wheel, I was unfamiliar with some key concepts from ecological design that could have been useful. For example, I might have incorporated more about a product's lifecycle into the section on material reality/environment and included more reading materials about sustainability design, such as cradle-to-cradle and biomimicry design (I did show short YouTube videos about these approaches).

Another issue is that the Ecomedia Wheel design lacks clear boundaries between some subject areas. For example, it was very difficult to distinguish between the environmental (material) aspects of the gadget from political economy. And it was challenging to separate political economy from culture. Though the point of the Ecomedia Wheel is to show how all these work together as an iterative process, I am hard-pressed to find exactly where these different areas of inquiry are distinct. They seem to bleed together quite a bit, which demonstrates why container metaphors that separate areas of inquiry are inadequate for holistic thinking and systems approaches.

Finally, the assignments reinforced some of the practices that I have critiqued in this book. For example, by stressing individual assignments, I was promoting the autonomous and isolated self. To remedy this, it would be good to create aspirational group assignments based on an essential question. To be fair, in a 14-week course I was constrained by time and the need to cover predetermined topics; however, with some creativity I think it's possible incorporate EfS and ecopedogogical strategies that involve action-oriented projects with creative problem solving as a guiding principles.

## Conclusion: Greening the Future of Media Literacy Education

During the case study, about halfway through the semester, I asked students to raise their hands if any of them expected the course to be about ecology. None did. Yet, by the end of the semester it was clear that many students understood why digital media and sustainability are connected. In setting out to redesign the course, it was important for me to avoid some of the problems I had encountered with traditional media-literacy approaches. Namely, I wanted students to contextualize inquiry according to their daily lives. I did not want to be overly didactic but to make the theories we discussed in the classroom relevant to their personal experience. I also wanted to bridge media analysis with media practice; typically media educators only do one or the other. Though students engaged in traditional course activities, such as reading texts and writing papers, they also used the medium they were studying, such as the course blog, YouTube and Prezi. I also feel they understood how their media usage is ecological. As such, I believe I achieved my objective to go beyond functionalist and protectionist media literacy by holistically combining critical, yet personally relevant, engagement with media technologies. Based on student feedback, personalizing the curriculum to make it practical and visceral created the space for them to link ecology with media.

The ecomedia literacy framework is still a work in progress. This example was in a university classroom where focused study is possible. In other environments where short-term units or activities are the norm, it might be more difficult to implement. It needs to be further tested to see how translatable the framework is for other courses and disciplines. For example, it would be interesting to see this approach applied in a setting that is exclusively dedicated to sustainability studies or integrated into business programs. In terms of lessons beyond greening an undergraduate class, I believe the ecomedia literacy framework is flexible enough to work in different contexts. In particular, I would like to see it experimented with in less formal education environments. Moreover, I would like to see it applied to situations that are not focused on gadgets, but instead on media texts or themes related to food, animals, energy, or public health. My experience in the media literacy movement, despite its problems, taught me that media education is a fun and engaging way to teach about social issues. Imagine if environmental educators could do the same by using media to teach about sustainability.

The model I offer here is experimental and not definitive. My main goal is to assert that ecological issues can be integrated into a media-studies course that typically eschews the environment. This experiment verified for me that a standard media-studies approach, and by extension media literacy, could be greened without compromising key disciplinary concepts. In fact, incorporating ecological themes strengthens the study of media because it expands the notions of democracy, social justice, and participation that are so important to media educators. Finally, the case study further reinforced for me that teaching green cultural citizenship should become a widely adopted goal for media educators and beyond. Based on student feedback, I believe they appreciated gaining an awareness of how to incorporate green cultural citizenship into their media usage.

CHAPTER SEVEN

# Media as Sustainability Education

The connection between media, culture, and technology has evolved far beyond its roots in mechanistic thinking. As such it is my hope that the current evolution of media can be harnessed for the purpose of sustainability. By moving beyond a 20th-century mass society framework, media practitioners can reintroduce the democratic, and hence participatory, potential of media that had been limited under a previous hierarchical media environment. It is also my hope that greater interconnectivity can promote green cultural citizenship and lead to new conditions for social change. This is certainly part of the current thinking about emerging social movements throughout the Arab world (Mason, 2012) and the rise of the "networked fourth estate" (Benkler, 2011). Nonetheless, I believe that a critical, if not agnostic, stance needs to be maintained, because the internet is not immune to the process of enclosure (the privatization of the cultural commons) and can also be used as a tool for government repression (Morozov, 2011).

As in media studies, most classic media literacy texts were written before the rise of social media, so most do not directly address the phenomena of Web 2.0. As discussed, media literacy tends to focus on how people interpret and respond to media content. However, in recent years there has been a rise in a new kind of literacy based on participation and collaboration, what Rheingold (2012) calls "networked social learning." Gauntlett (2011) grounds this phenomenon in the tradition of the do-it-yourself maker movement, arguing that craft, creativity, and community are outcomes of a "making-and-doing culture" happening in and outside the web. In contrast to traditional, mechanistic forms of media literacy, it can be argued that this kind of learning self-consciously represents the truest example of McLuhan's aphorism that the medium is the message. Here, learning is specifically tool oriented and is based on practices engendered by these tools. It does not emphasize the analysis of messages advocated by media literacy educators. Rather, Guantlett proposes that mediamaking offers pleasure, enabling learners to see themselves as creative agents dialoguing within a larger community of practitioners. They do this in the hope of recognition by and connectivity with others.

Along these lines, the open education movement advocates for education without barriers, leveraging the strength of open systems enabled by networked media. Examples of open education include MIT's OpenCourseWare, OER Commons (Open Education Resources), and P2P University, which advocate learning that is informal, self-directed, and lifelong. These approaches are inspired by emerging literacies, such as participatory cultural practices (De Abreu, 2011; D. Thomas & Brown, 2011), informal learning (Cross, 2007), digital storytelling (Alexander, 2011), design literacy (Sheridan & Rowsell, 2010), and online communities of practice (Beetham & Sharpe, 2007; Downes, 2007; Gurell, Kuo, & Walker, 2010; Harlen & Doubler, 2007; Weller, 2002; Wenger, White, & Smith, 2009). A related development to emerge in the area of social media and education is a new pedagogical model referred to as "connected learning" (http://connectedlearning.tv). Advanced under the auspices of the MacArthur Foundation, it promotes the following components: learning principles that are interest powered, peer-supported, and academically oriented; design principles that are production centered, openly networked, and have a shared purpose; and core values based on equity, social connection, and full participation. Though the connected learning community has yet to incorporate a sustainability perspective into its approach, I see a lot of potential for this method to correlate with green cultural citizenship, since it emphasizes design principles, participation, and relationships.

The rise of social media education practices fulfills the inspired vision of Ivan Illich's (1971) deschooled society, which emphasizes peer-based, lifelong, and informal learning. This is based on Illich's differentiation between networks that are open (*learning webs*) and closed (*manipulative institutions*). Illich believed that schools are an expensive means for conditioning people to learn how to be institutionalized. In such an environment, learning can only be informed by experts and paid professionals, reinforcing dependence on the irrational inner-logic of bureaucracy without regard to that which is practical in daily life. This is best exemplified by the disconnection between standardized testing and those skills necessary to deal with the ecological crisis. If we apply agricultural metaphors, Illich's model of education can be viewed as a kind of permaculture, whereas formal education is a kind of monocultural crop production.

## Towards a Healthy Media Ecosystem

As do members of any living ecosystem, we live and breathe the environment we are part of. But as humans, we also have cultural and social structures that impel us to be self-conscious and aware that we have rights and responsibilities to maintain this system. This prompts an extremely important guiding question: What form does a healthy media ecosystem take? Ultimately this question underlies the purpose of green cultural citizenship, necessitating a positive vision for the future. But like language, the media ecosystem is collectively created, so a single proposal imposes upon the system's creative capacity for collective and emergent solutions. Nonetheless, I see glimpses of it currently prototyped by a vast array of seemingly disconnected groups and practitioners, yet as a whole they appear to be working towards an alternative future, one that Paul Hawken (2007) calls the largest social justice and environmental movement in human history. No doubt, media play a significant role by enabling these groups to share ideas, inspire each other, and to coordinate. Of particular interest are social movements, such as Spain's *indignatos* (the indignants), Arab Spring, and Occupy Wall Street; creative artists and storytellers; and climate disruption and sustainability activists. These three areas (social movements, storytelling, and sustainability communication) suggest that media can indeed become a kind of sustainability education because built into their practices are a kind of pedagogy that advocates the change they want to become. Through practice they produce literacies of sharing, cooperation, togetherness, and participation that engender trust. All of these activities constitute current practices within the media ecosystem, and if it is true that all members are designers of meaning, then we can all become sustainability designers.

**Social Movements**

For several decades, Manuel Castells has studied extensively the changes resulting from our interactions on the internet, what he calls the *network society* (Castells, 1996, 1997, 2000). His most recent work ( 2011, 2012) updates how these changes are being expressed through various social movements around the world. One of his key insights is that we have encountered a shift from mass media to mass self-mediation through our social networks. Particularly with young people, this is leading to a shift in values that embrace *togetherness*, *sharing*, and *autonomy*. For Castells, togetherness is a desire to work collectively and is used instead of community, which implies more long-term and stable relationships. Autonomy is differentiated from individualism and is defined as working independently from traditional institutions in hybrid spaces that are online and in public, urban spaces. These groups work leaderless and horizontally in open-ended networks, and within networks of networks, to promote social change. Castells (2012) writes, "the role of the Internet goes beyond instrumentality: it creates the conditions for a form of shared practice that allows a leaderless movement to survive, deliberate, coordinate and expand" (p. 229). Furthermore, he contends "the Internet provides the organizational communication platform to translate the culture of freedom into the practice of autonomy" (p. 231).

An example of this is a group of media activists in Cairo, Mosireen (http://mosireen.org), which created an alternative media collective where activists can gather and share information about the Egyptian uprising; store and maintain archives of citizen media; and produce outdoor screenings in public spaces of their media work. Meanwhile, this media is available publicly on the internet. Another example is Outta Your Backpack Media (http://oybm.org), a Native American media activist collective that produces documentaries about environmental issues and other concerns of Native American youth. The collective's name comes from a project they developed that fits a mobile mediamaking system in a rucksack that can be ported around by skateboard. This group also runs media literacy workshops.

According to BBC reporter and economics editor Paul Mason (2012), many of these social movements are using media creatively and for different purposes according to the medium. For example, Twitter supplies information updates; social media (Facebook, etc.) enables ad hoc groups to form and coordinate; video sites (YouTube, Vimeo) and photo-sharing services (Flickr, Picasa) provide evidence; blogs produce analysis, research, linking, and context; and traditional broadcast media (such as Al Jazeera) reach people who are offline while drawing on information from online networks. Also emerging from these various information-sharing processes has been the phenomena of news "curators" who are experts or knowledgeable informants who monitor the networks to filter and select what is important (Carvin, 2012). In my Media Ethics course, I have students use curation platforms such as Scoop.it (http://scoop.it) and Storify (https://storify.com) to report on media ethics scandals. By doing so, they have to incorporate their own ethical practices and also learn how to communicate effectively. This is reinforced by Storify's tagline, "Make the web tell a story," which hints at another important literacy afforded by new media practices: digital storytelling.

## Storytelling

Jonah Sachs, a co-founder and CEO of Free Ranges Studios, helped produced the viral web video, *The Story of Stuff* (Story of Stuff Project, 2009). In his book, *Winning the Story Wars: Why Those Who Tell (and Live) the Best Stories Will Win the Future* (Sachs, 2012), he uses ecological metaphors to describe the significance of storytelling in the 21st century:

> For all of its high-priced barriers to entry, the broadcast tradition makes life reassuringly simple for marketers. By artificially selecting what gets published, it goes a long way toward ordering the world's ecosystem of ideas into something resembling a modern industrial farm—monocultured, predictable, controlled. Ideas in the oral tradition, on the other hand, look like the wild "entangled bank" that Charles Darwin contemplated in *The Origin of Species*—chaotic, seething, freed from artificial selection. Here survival of the fittest rules. In the natural world, memes, or ideas, must ensure their survival by exciting listeners to keep passing them along, carrying the same core message in a chain of transmission. (pp. 16-17)

Though I disagree with his mechanistic approach to communication ("carrying the same core message in a chain of transmission"), I do agree with Sachs' assertion that the stories that succeed in our complex media environment are the stories that matter. In a sense, this is what emerged during the last several years with global social movements, whose mediamaking practices countered stereotypes and promoted empathy, solidarity, connectivity, ownership, and empowerment. Stories that express these values lead to an emotional literacy that is an antidote to the superficial emotions generated by commercial media and marketing. Moreover, it suggests that media can have a regenerative power that supports resilience, a key stance for sustainable cultural practice (Edwards, 2010).

Furthermore, digital storytelling, which includes gaming, remixing, blogging, podcasting, multimedia, augmented reality, web video making, and social networking (Alexander, 2011) can support M. C. Bateson's (2007) call for narratives of connection. Stories are present in popular culture, such as *Avatar* (Cameron & Landau, 2009), and in grassroots media such as *The Story of Stuff* (Story of Stuff Project, 2009). A beautiful example is an online video, *Overview* (Reid & Ferstad, 2013), which tells the story of the quasi-religious epiphanies of astronauts who see earth from space. Although politically and ethically flawed, *Kony 2012* (Invisible Children, 2012) was an extremely well-executed viral video campaign that ignited activism targeting Ugandan warlord Joseph Kony. These examples and many more provide inspiration for sustainability mediamaking.

My own personal experience has taught me that storytelling goes hand in hand with critical media literacy, which combines deconstruction with reconstruction (López, 2011c). Students can learn from marketers the basic techniques of effective communication and then reinterpret those techniques to tell new stories. I have done this several times in Native American communities where students first learned basic media-literacy skills and then went on to produce their own videos. These projects culminated in community screenings where students shared their work with authentic audiences. Sustainability advocates could do similar projects that promote regenerating the commons.

**Sustainability Communication**

Greenpeace, 350.org, and anti-fracking activists are all using media to make claims and to shift the debate about key environmental issues. Bill McKibben (2008) argues for the media equivalent of a farmers market in which conversation is encouraged. Community radio best fits this model, but there are many other kinds of media that sustainability activists are creating. Greenpeace uses online videogames, viral video, and stages *image events* to get media attention to the causes it supports. 350.org has used many different media campaigns, such as "Connect the Dots," information graphics, and viral videos to spread its message and to organize and coordinate activism around the country and on college campuses. Josh Fox's anti-fracking documentary *Gasland* (Adlesic, Gandour, Fox, & Roma, 2010) has inspired an assortment of grassroots media campaigns, including Sean Lennon and Yoko Ono's "Don't Frack My Mother" (http://artistsagainstfracking.com/dont-frack-my-mother).

## Small Media Are Beautiful

The notion of catastrophic global ecological collapse is often too abstract and large for the average person's immediate horizon line, but that doesn't mean it should be left to the "knowledge experts" to set the local agenda. Recalling E.F. Schumacher's (1993) dictum that "man is small...therefore small is beautiful," we could benefit from a scaling down of our vision to reorient ourselves to a landscape perspective (while also holding a space for the global dimensions of our current reality) in order to strengthen the cultural commons and promote intergenerational dialogue within an inhabitable scale of perception. As Robert J. Brulle and J. Craig Jenkins (2006) argue,

> Movements are more effective if they engage citizens in a sustained dialogue rather than treating them as mass opinion to be manipulated.... Is the cure to create new spin-doctors who promote different unified progressive frames? Or is it better to generate a genuine dialogue that creates value change and better understanding of both self and public interest? (p. 85)

Likewise, Schumacher proposes, "The case for hope rests on the fact that ordinary people are often able to take a wider view, and a more 'humanistic' view, than is normally being taken by experts" (1993, p. 130).

As Gustavo Esteva and Madhu Suri Prakash (1998) are right to point out, "global thinking" is an oxymoron because we can never know how another culture outside of our own thinks:

> Excluded...from critical scrutiny is the reflection that in order for "global thinking" to be feasible, we should "think" from within every culture on Earth and come away from this excursion single-minded—clearly a logical and practical impossibility, once it is critically de-mythologized. For it requires the supra-cultural criteria of assuming that it is possible to "think" outside of the culture in which every man and woman on Earth is immersed. The human condition does not allow such operations. We [the authors] celebrate the hopefulness of common men and women, saved from the hubris of "scientific man," unchastened by all his failures at playing God. (p. 23)

Esteva and Prakash critique the kind of sweeping, abstract thinking characteristic of Western technological thought, activism, and education. The location of resistance, they argue, is really in the person's daily actions when he or she engages (or disengages, as the case may be) multinational corporations or governments doing their bidding. A group like the Zapatistas in Chiapas, Mexico, for example, is a local response to the global neoliberal project. Though its cause resonates with other international struggles, its goals and activities are geared towards strengthening the community's response to the radical ideology of global capital. The fact is everyone lives somewhere and is in contact with the causes of the crisis. To ignore the local dimension of biodiverse intelligence is to deny a multicultural response to the situation.

By locking ourselves into a monocultural media-literacy strategy, we also deny ourselves a humanistic, and hence truly ecological and transformative, approach to the problem. In her discussion of sociologist Philip Slater's work, ecopsychologist Chellis Glendinning (1994, p. 94) points to three "universal urges" that are "frustrated" in the blind strategy of our mass mediated society: "the urge for community, the urge for engagement, and the urge for shared responsibility and interdependence." The bridge to these "urges," I would suggest, starts with authenticity, which is constantly disrupted by our postmodern media and marketing in which one cannot tell a sales pitch from a genuine conversation. Subsequently, we need to consider *small* and *slow* media. By "small" I don't mean Twitter because though it is a kind of micro-communication, it is not slow, and its size is due to mechanized fragmentation as opposed to human-scaled, live feedback conversation. What I have in mind is the organic media equivalent of slow food, which means that meals are not consumed like fast food, but done ritualistically by groups of people who cook with fresh ingredients that are produced bioregionally. To bring it back to Carey (2009), again we are talking about the difference between "transmission" and "ritual" (see Chapter 2). Bill McKibben (2008, p. 131) alludes to the loss of a local identity when he compares box-store economics with mass media: "And like food radio used to be mostly local, hemmed in by mountains, limited by signal strength...for the most part radio served a *place*..." He further laments,

> We are the most comprehensively entertained people in history. Between our hundreds of channels of television, and radio, and Internet radio, and our legally and illegally downloaded tunes, there is no vaguely musical sound emitted by anyone on the planet that is not available (for a vanishingly low price) at any time of the day or night. But oddly, it's gotten a lot harder to hear much about your immediate vicinity." (p. 133)

Calling for a "sonic farmers market," he points to studies that show in a farmer's market you are ten times more likely to have a conversation than in a supermarket. Likewise, a community radio station is more likely to generate discussion and dialogue about what matters in a community than satellite-beamed content streaming disembodied talking heads who are ticking off bullet-point cognitive frames.

Promoting community media (and hence small and slow media) in response to the dangers of escalating climate chaos feels good, but defining community is more problematic. "Community" is one of those words that universally sounds great but is tricky to define. First it's necessary to say what it is not: demographics. From insurance companies to ad rate cards, companies compile stats to demarcate and predict people's behaviors. Our digital profiles on online social networks consist of interests like books, bands, films, and TV shows (all cultural commodities) as signs of who we are and where we belong. We dress certain ways and buy into brands to further identify our loyalties (Mac vs. PC, Volkswagen vs. Hummer, etc.), yet if you press people to define what it is that makes them a part of a community, a different set of expressions take root: connectedness, shared experience, values, history—in short a common sense of meaning. We can also experience such affiliations as members of subcultures—like punk, hip hop, emo, or goth—but a community still has something that demographics and style usually don't have: a sense of place. Even if it is the most decentered locale in the world, like Los Angeles, there is still a sense of being from there that conditions and reveals much about one's sense of self. These kinds of ties are not meaningless and deserve strengthening in response to the depersonalization and utilitarian uses of people's human expressions as markets.

In this context, Carey (2009) has an extremely valuable definition of communication: "communication is a symbolic process whereby reality is produced, maintained, repaired, and transformed" (p. 19). I want to emphasize the "repaired" and "transformed" part of the equation, because that reintroduces the social aspects of communication in which we work together to resolve contradictions in our communities and economic practices. As Etiene Wenger (1998) discusses regarding his theory of communities of practice,

> Effective communication or good design...are not best understood as the literal transmission of meaning. It is useless to try to excise all ambiguity; it is more productive to look for social arrangements that put history and ambiguity to work. The real problem of communication and design then is to situate ambiguity in the context of a history of mutual engagement that is rich enough to yield an opportunity for negotiation. (p. 84)

As we know from participating in any conversation, ambiguity is always the space that requires some kind of negotiation and feedback. The beauty of organic media is that it is alive, and it embraces uncertainty because it is done on a human scale.

A community radio station, media literacy class, or a discussion group around an independently made documentary would be along the lines of Schumacher's idea of "intermediate technology." In Schumacher's following remark, just substitute the word "media" for "technology": "One can also call it self-help technology, or democratic or people's technology—technology to which everybody can gain admittance and which is not reserved to those already rich and powerful" (1993, p. 127). Slow media can be poetry slams or community film festivals that draw in local businesses and community members. One example is the Food for Thought Film Festival, which was a film series in New York City's underserved communities that invited organic farmers, caterers, and vendors to attend and exhibit at the festival. Robert Greenwald's Brave New Films' (http://bravenewfilms.org) model of house-party screenings and word-of-mouth distribution offers another approach. Then there is the Taos Solar Music Festival (www.solarmusicfest.com), a gathering for music and environmental education. In this particular form of slow media, dance and celebration are key aspects of revitalizing our ecological selves, which have been so dehumanized by all the large-scale technological projects of civilization. Bioneers deliberately stays small, knowing that becoming too big is counter to its mission. It has created various satellite hubs to accommodate its growing popularity, to keep the discussion on a bioregional scale.

Then there are the possibilities of the internet to facilitate communities of practice, reassert a creative and cultural commons, and to be an organizational tool for people helping each other and to make connections: CouchSurfer.com, MeetUp.com, and craigslist.com (which spontaneously coordinated aid for New Orleans after Katrina) are just a few examples of how human connectivity is facilitated by websites. Slow and small internet activities are examples of an organic media paradigm that aims to promote a cultural commons, something that is constantly under attack and dismantled in the process of capitalist enclosure. It's a cultural immune system response. Organic media should follow important architectural considerations, elements already present in the emergent practices of new media users, such as open source, sharing protocols, and Creative Commons, all of which are vital characteristics of living and open ecological systems.

As stated, media education addresses involvement in Silverstone's (2007) concept of the mediapolis, our collective space of civic and democratic participation within the realm of globally mediated technology. He asserts citizenship skills require "the kinds of judgment and reflexivity which underpin the idea of literacy at every level in the mediapolis: a critical understanding of media and mediation as a global practice with significant consequences for the way we all live" (2007, p. 182). Mediapolis is a useful metaphor for media, for it invites comparisons with place, in particular, a cosmopolitan city and the discussion of the "right to the city":

> The right to the city is far more than the individual liberty to access urban resources: it is a right to change ourselves by changing the city. It is, moreover, a common rather than an individual right since this transformation inevitably depends upon the exercise of a collective power to reshape the processes of urbanization. The freedom to make and remake our cities and ourselves is, I want to argue, one of the most precious yet most neglected of our human rights. (Harvey, 2008, p. 23)

If learners envision themselves as inhabiting media as a kind of place—in the same sense of the city as a space of civic engagement with a material reality—then it may be possible to move beyond a model of media education that views media as purely a system of symbolic processes dominated by corporations. Like inhabitants of the city, in a mediapolis we are not merely participants, users, or prosumers, but members who are part of communities with rights and responsibilities. Cities also require sustainability design and planning. What if these concepts were applied to media?

By augmenting the definition of who and what is part of the media ecosystem, media education ceases to be a matter of studying software, gadgets, messages, corporate design, and government policy. It is broadened to include laborers in African mines that extract rare earth minerals, workers in Chinese factories, and electronic rag pickers who extract metals from e-waste (Maxwell & Miller, 2012). From an ethical framework that incorporates the rights of biotic communities (Leopold, 1987), we would have to consider the membership of living systems, such as waterways impacted by chemicals, or animal habitats destroyed by mining or soil system depletion. Such a framework entails advocacy for a particular vision of democracy and the public sphere based on the cultural commons and Earth Democracy, which is based on the Indian notion of *vasudhaiva kutumbakam*, earth family (Shiva, 2005). Earth family encompasses the planetary community of beings that comprise our living systems and should be considered part of an ecocentric vision of democracy.

Throughout this book I have asserted that media and education are similar in that they are examples of meaning design. That is, each creates a context that generates value. Furthermore, I have suggested that whether conscious or not, both media and media literacy are a kind of environmental education. By showing how mechanism pervades media and media literacy, I have mostly discussed this in the negative sense. However, I believe certain kinds of media can be viewed as sustainability education that promotes green cultural citizenship. In other words, new media practices are themselves pedagogical, promoting new social values and norms that transcend the old approach of teaching about media. As a result, by participating as members of the media ecosystem, rather than learn about the media, we learn how to inhabit the media.

## Coda

In a 1991 article, Kathleen Tyner, a veteran (and sometimes referee) of ongoing media literacy debates, used the parable of the blind men and the elephant to describe the state of media literacy debates. In this Indian folktale, several blind men grope parts of an elephant to determine what it is. Each one independently is unable to imagine the elephant, but instead envisions something entirely different. Tyner wrote, "Media educators in the United States are a fractious bunch. One teacher's definition of media education is another's heresy. Like the blind men and the elephant, teachers often practice one small aspect of media education and conclude that they have the whole picture" (Tyner, 1991, para. 1). Twenty years later at the Media Education Manifestos website, Tyner wrote again about the state of the field. She suggests that media educators have a propensity to search out taxonomies based on their own bias to favor textual analysis and deconstruction as professional practice. But in our efforts to narrowly define the particular debates and to struggle over disciplinary boundaries, we also may be losing sight of a much larger issue:

> Although the link between communication and culture is obvious to media educators, it is still a struggle to think past the assumptions of my own literacy patterns and to engage with the way that literacy reinforces ingrained social norms, power structures and pathways to social capital across cultures. In policy circles, media educators still tend to focus on media education definitions, standards, assessments, key concepts and curricula. These discussions are useful as snapshots of literacy in motion, although occasionally they play out as reiterative and tiresome turf battles. (Why do scholars always have to name everything?) Rarely do they provide media educators with the cultural distance necessary to glimpse the whole array of literacy affordances (and limitations)—the sum of its parts. Instead, the focus on deconstruction, definitions and details may make it even more difficult to engage with the innovative and creative educational practices that connect us to contemporary literacy skills and practices. (Tyner, 2011, para. 1)

Tyner's observations imply that some form of combined observations can help synthesize the disparate debates and struggles to offer scenarios of "media literacy in motion." I do not claim to be any smarter than a single blind man/woman touching the media literacy elephant, but I do feel that my analysis at least offers a perspective that has yet to be considered. For example, the view of media literacy as an ecosystem is a novel approach that allows for the generation of new ideas about the field.

In this sense, I believe that my exploration of the media literacy ecosystem is an effort to create some distance (albeit biased with a different set of assumptions) to identify the cultural norms that underlay the boundary-making discourses of practitioners. Examining each site of analysis was like exploring one blind person's interpretation of the media literacy elephant; individually each site offered a perspective, but taken as a whole a larger pattern emerged. Given the diversity of the media literacy ecosystem, it should be obvious that there is no single figured world that represents a prototypical *media literacy educator*, yet by engaging in boundary work it is possible to make visible those boundaries that define the field.

Orr's (1994) suggestion that all education is ecological education challenges us to consider that teaching and learning is not just about the transfer of information but is an encounter of worldviews. Parallel to media literacy education's evolution as an international movement is a growing planetary emergency in the form of an ecological crisis that threatens to undermine civilization. Unsustainable levels of $CO_2$ in the atmosphere, the acidification of the oceans, and the loss of biodiversity are key issues contextualizing the world in which media education is formulated. Ultimately, a failure to address the ecological crisis undermines the aims of education, which can be viewed as the method by which culture sustains itself. To put it bluntly, the editors of *Rethinking Schools* (2011) declared, "our climate crisis is an education crisis." Moreover, we are faced with an ongoing economic emergency, which is closely tied to our ecological predicament (Gilding, 2012; Korten, 2006). Indeed, the economic and ecological crises that we now face demand new life skills that will help reinvent how we engage the world. Subsequently, because most of our educational programs were designed under previous social conditions, current assumptions about the nature of education and learning are being confronted by an assortment of challenges, including the disassociation between the mediated world and the classroom experience (Gutiérrez-Martín & Tyner, 2012). Following the pattern of a dying paradigm described by Kuhn (1996), mainstream education's reductive and hierarchical structures inherited from 19th-century pedagogical models are breaking down (RSA, 2010). Efforts to bring ecologically oriented cultural practice into the realm of media education are tied to how education is being re-thought of in the context of so-called 21st-century skills, which, among many proficiencies, promotes systems thinking, problem solving, collaborative learning, critical thinking, lifelong learning, and digital media literacy. In this context, "The concept of what a school is does not need *reform*—it needs *new* forms" (Jacobs, 2010, p. 9, emphasis original). Thus, we are in a transitional moment that is uncertain but also hopeful because from crisis also comes innovation.

Likewise, after a century of media theory, we should be well beyond a content-only approach to communications. If we are to bridge green cultural citizenship with media literacy, it would be wise to consider the pedagogical dimension of media and to explore to what extent they have ecological worldviews embedded into them. Thus, when it comes to devising a response to planetary ecological disruption, it is my contention that we need to heed Berry's (2005) "design for pattern approach," which means developing a healthy, organic media alternative to pre-existing mass communication strategies that tend to speak in the same language as the Industrial and

Scientific Revolutions. As a media literacy educator who has worked in diverse, multicultural environments, I have learned that we must find new communication models that do not rely on the same modes of thought that brought us to the current crisis. In particular, we need to encourage models that respect biocultural voices in a fluid and participatory manner. I suggest that a *shallow* method of media literacy is mechanistic, akin to Shiva's (1993) concept of monocultural thought. By contrast I propose a *deep* method that is designed for open and local contexts. As such, I believe ecomedia literacy should evolve *slow* and *small* education environments that encourage green cultural citizenship.

By applying ecological metaphors, we can reconceive the meaning design of education as it relates to media and propose new ways of inquiring and engaging media so as to promote healthy and sustainable behaviors. My vision for the future of media literacy education is that green subjects are not "ghettoized" and treated as distinct or off topic from those subjects that are familiar to us. I imagine that all media education one day will incorporate sustainability, and it will be "natural" to do so. I believe the barriers can be overcome, but it will take concerted effort and will be up to the practitioners (i.e., teachers, scholars, policymakers, activists, and learners) to push for more integrative approaches to teaching media.

# Bibliography

Abercrombie, N., & Longhurst, B. (2007). *The Penguin dictionary of media studies*. London: Penguin Books.

Abram, D. (1996). *The spell of the sensuous: Perception and language in a more-than-human world*. New York: Pantheon Books.

Adlesic, T., Gandour, M., Fox, J., & Roma, D. (Producers), & Fox, J. (Director). (2010). *Gasland* [Motion picture]. United States: New Video Group.

Alakeson, V. (2003). *Making the net work: Sustainable development in a digital society*. Middlesex, U.K.: Xeris.

Alexander, B. (2011). *The new digital storytelling: Creating narratives with new media*. Santa Barbara, CA: Praeger.

Allen, T. F. H., Tainter, J. A., & Hoekstra, T. W. (2003). *Supply-side sustainability*. New York: Columbia University Press.

Altheide, D. L. (1995). *An ecology of communication: Cultural formats of control*. New York: Aldine de Gruyter.

Altheide, D. L. (1996). *Qualitative media analysis*. Thousand Oaks, CA: Sage.

Altheide, D. L., Coyle, M., DeVriese, K., & Schneider, C. (2010). Emergent qualitative document analysis. In S. N. Hesse-Biber, & P. Leavy (Eds.), *Handbook of emergent methods* (pp. 127–151). New York: Guilford Press.

Anderson, A. (1997). *Media, culture, and the environment*. London: UCL Press.

Anderson, B. (1983). *Imagined communities: Reflections on the origin and spread of nationalism*. London: Verso.

Aufderheide, P. (1993). *Media literacy. A report of the National Leadership Conference on Media Literacy*. Washington, DC: Aspen Institute, Communications and Society Program. Retrieved from http://www.eric.ed.gov/ERICWebPortal/detail?accno=ED365294

Bagdikian, B. H. (2004). *The new media monopoly*. Boston, MA: Beacon Press.

Baichwal, J. (Director). (2006). *Manufactured landscapes* [Motion picture]. Canada: Foundry Films.

Bakhtin, M. M., & Holquist, M. (1981). *The dialogic imagination: Four essays*. Austin: University of Texas Press.

Banerjee, S. C., & Kubey, R. (2014). Boom or boomerang: A critical review of evidence documenting media literacy efficacy. In E. Scharrer, A. N. Valdivia, J. C.

Nerone, V. Mayer, S. R. Mazzarella, R. E. Parameswaran, & K. Gates (Eds.), *The international encyclopedia of media studies*. Malden, MA: Wiley-Blackwell.

Baran, S. J. (2004). *Introduction to mass communication: Media literacy and culture*. Boston: McGraw-Hill.

Barber, B. R. (2006). Pangloss, Pandora or Jefferson? Three scenarios for the future of technology and strong democracy. In R. Hassan, & J. Thomas (Eds.), *The new media theory reader* (pp. 188-202). New York: Open University Press.

Barker, C., & Galasiński, D. (2001). *Cultural studies and discourse analysis: A dialogue on language and identity*. Thousand Oaks, CA: Sage.

Barlow, J. P. (1993). Selling wine without bottles: The economy of mind on the global net. Retrieved from https://w2.eff.org/Misc/Publications/John_Perry_Barlow/HTML/idea_economy_article.html

Barnosky, A. D., Hadly, E. A., Bascompte, J., Berlow, E. L., Brown, J. H., Fortelius, M., Smith, A. B. (2012). Approaching a state shift in earth's biosphere. *Nature, 486* (7401), 52-58. doi:10.1038/nature11018

Barry, A. (2006). On interactivity. In R. Hassan, & J. Thomas (Eds.), *The new media theory reader* (pp. 163-187). New York: Open University Press.

Barry, J. (1999). *Environment and social theory*. New York: Routledge.

Bateson, G. (2000). *Steps to an ecology of mind*. Chicago: University of Chicago Press.

Bateson, M. C. (2007). Education for global responsibility. In S. C. Moser, & L. Dilling (Eds.), *Creating a climate for change: Communicating climate change and facilitating social change* (pp. 281-291). Cambridge, U.K.: Cambridge University Press.

Beder, S. (1998). *Global spin: The corporate assault on environmentalism*. Cambridge, U.K.: Chelsea Green.

Beetham, H., & Sharpe, R. (2007). *Rethinking pedagogy for a digital age: Designing and delivering e-learning*. London, U.K.: Routledge.

Benjamin, W. (1970, July-August). Author as producer. *New Left Review, 1*(62).

Benkler, Y. (2006). *The wealth of networks: How social production transforms markets and freedom*. New Haven, CT: Yale University Press.

Benkler, Y. (2011). A free irresponsible press: WikiLeaks and the battle over the soul of the networked fourth estate. Retrieved from http://benkler.org/Benkler_Wikileaks_current.pdf

Bennett, P., Kendall, A., & McDougall, J. (2011). *After the media: Culture and identity in the 21st century*. London, U.K.: Routledge.

Bentz, V. M., & Shapiro, J. J. (1998). *Mindful inquiry in social research*. Thousand Oaks, CA: Sage.

Berger, J. (1973). *Ways of seeing*. New York: Viking Press.

Berkes, F. (1999). *Sacred ecology: Traditional ecological knowledge and resource management*. Philadelphia, PA: Taylor & Francis.

# Bibliography

Berry, W. (2005). Solving for pattern. In M. K. Stone, & Z. Barlow (Eds.), *Ecological literacy: Educating our children for a sustainable world* (pp. 241-249). San Francisco, CA: Sierra Club Books.

Bioneers. (2013). What is bioneers? Retrieved from http://www.bioneers.org/what-is-bioneers/

Blewitt, J. (2006). *The ecology of learning: Sustainability, lifelong learning, and everyday life.* Sterling, VA: Earthscan.

Blewitt, J. (2009). New media literacy: Communication for sustainability. In A. Stibbe (Ed.), *The handbook of sustainability literacy: Skills for a changing world* (pp. 111-116). Totnes, U.K.: Green Books.

Blewitt, J. (2010). Deschooling society? A lifelong learning network for sustainable communities, urban regeneration and environmental technologies. *Sustainability, 2*, 3465-3478. doi:10.3390/su2113465

Blewitt, J. (forthcoming). Public libraries and the right to the [smart] city. *International Journal of Social Ecology and Sustainable Development*.

Bowers, C. A. (2000). *Let them eat data: How computers affect education, cultural diversity, and the prospects of ecological sustainability.* Athens, GA: University of Georgia Press.

Bowers, C. A. (2008). *Toward a post-industrial consciousness: Understanding the linguistic basis of ecologically sustainable educational reforms.* Eugene, OR: Eco-Justice Press.

Bowers, C. A. (2009). The language of ecological intelligence. *Language & Ecology, 3*(1), 1-24.

Bowers, C. A. (2012). *The way forward: Educational reforms that focus on the cultural commons and the linguistic roots of the ecological/cultural crises.* Eugene, OR: Eco-Justice Press.

Boyd-Barrett, O. (2002). Theory in media research. In C. Newbold, O. Boyd-Barrett, & H. v. d. Bulck (Eds.), *The media book* (pp. 1-54) London: Arnold.

Boyle, J. (1997). A politics of intellectual property: Environmentalism for the net? *Duke Law Journal, 47*(87), 86-116.

Boyle, J. (2008). *The public domain: Enclosing the commons of the mind.* New Haven, CT: Yale University Press.

Brereton, P. (2005). *Hollywood utopia ecology in contemporary American cinema.* Bristol, U.K.: Intellect.

Briggs, M. (2005). Rethinking school lunch. In M. K. Stone, & Z. Barlow (Eds.), *Ecological literacy: Educating our children for a sustainable world* (pp. 241-249). San Francisco, CA: Sierra Club Books.

Brulle, R. J., & Jenkins, J. C. (2006). Spinning our way to sustainability. *Organization & Environment, 19*(1), 82-87.

Brulle, R. J., & Young, L. E. (2007). Advertising and individual consumption levels 1900-2000. *Sociological Inquiry, 77*(4), 522-542.

Buckingham, D. (1991). Teaching about the media. In D. Lusted (Ed.), *The media studies book: A guide for teachers* (pp. 12-35). London: Routledge.

Buckingham, D. (2003). *Media education: Literacy, learning and contemporary culture.* Cambridge, U.K.: Polity.

Buckingham, D. (2007). *Beyond technology: Children's learning in the age of digital culture.* Cambridge: Polity.

Buell, L. (2001). *Writing for an endangered world: Literature, culture, and environment in the U.S. and beyond.* Cambridge, MA: Belknap Press of Harvard University Press.

Cajete, G. (1994). *Look to the mountain: An ecology of indigenous education.* Durango, CO: Kivaki Press.

Cameron, J., & Landau, J. (Producers), & Cameron, J. (Director). (2009). *Avatar* [Motion picture]. Australia: Twentieth Century Fox.

Campaign, C. (2005). *Assessing communication rights: A handbook.* London, U.K.: World Association for Christian Communication.

Campbell, R. (2009). *Media and culture.* Boston, MA: Bedford/St. Martin's.

Cantrill, J. G., & Oravec, C. L. (Eds.). (1996). *The symbolic earth: Discourse and our creation of the environment.* Lexington, KY: University Press of Kentucky.

Capra, F. (1983). *The turning point: Science, society, and the rising culture.* Toronto, Canada: Bantam Books.

Capra, F. (1996). *The web of life: A new scientific understanding of living systems.* New York: Anchor Books.

Capra, F. (2004). *The hidden connections: Integrating the biological, cognitive, and social dimensions of life into a science of sustainability* . New York: Anchor Books.

Capra, F. (2005). Speaking nature's language: Principles for sustainability. In M. K. Stone, & Z. Barlow (Eds.), *Ecological literacy: Educating our children for a sustainable world* (pp. 18-29). San Francisco, CA: Sierra Club Books.

Capra, F. (2008). Systems theory and the new paradigm. In C. Merchant (Ed.), *Ecology* (2nd ed., pp. 365-372). Atlantic Highlands, NJ: Humanity Books.

Carey, J. T. (2006). Technology and ideology: The case of the telegraph. In R. Hassan, & J. Thomas (Eds.), *The new media theory reader* (pp. 225-243). New York: Open University Press.

Carey, J. T. (2009). *Communication as culture, revised edition: Essays on media and society.* New York: Routledge.

Carvin, A. (2012). *Distant witness: Social media, the Arab Spring and a journalism revolution.* New York: CUNY Journalism Press.

Castells, M. (1996). *The rise of the network society.* Malden, MA: Blackwell.

Castells, M. (1997). *The power of identity.* Malden, MA: Blackwell.

Castells, M. (2000). *End of millennium.* Malden, MA: Blackwell.

Castells, M. (2011). *Communication power.* Oxford, U.K.: Oxford University Press.

Castells, M. (2012). *Networks of outrage and hope: Social movements in the internet age.* Cambridge, U.K. & Malden, MA: Polity.

Cazden, C., Cope, B., Fairclough, N., Gee, J. et al. (1996, Spring). A pedagogy of multiliteracies: Designing social futures. *Harvard Educational Review, 66*(1), 60-92.

Chalquist, C. (2010). *The brain as archetypal tree (and other neurological nature metaphors).* Walnut Creek, CA: World Soul.

Charmaz, K. (2006). *Constructing grounded theory: A practical guide through qualitative analysis.* Thousand Oaks, CA: Sage.

Charmaz, K. (2010). Grounded theory methods in social justice research. In N. K. Denzin & Y. S. Lincoln (Eds.), *The Sage handbook of qualitative research* (pp. 363-397). Thousand Oaks, CA: Sage.

Christakis, N. A., & Fowler, J. H. (2011). *Connected: The surprising power of our social networks and how they shape our lives–how your friends' friends' friends affect everything you feel, think, and do.* New York: Back Bay Books.

Clarke, A. (2005). *Situational analysis: Grounded theory after the postmodern turn.* Thousand Oaks, CA: Sage.

Clarke, A., & Friese, C. (2010). Grounded theorizing using situational analysis. In A. Bryant, & K. Charmaz (Eds.), *The Sage handbook of grounded theory* (pp. 363-397). Thousand Oaks, CA: Sage.

Cloud, J. P. (2010). Educating for a sustainable future. In H. H. Jacobs (Ed.), *Curriculum 21: Essential education for a changing world* (pp. 168-85). Alexandria, VA: Association for Supervision and Curriculum Development.

Cloud Institute for Sustainable Education. (2011). *EfS curriculum design workbook.* New York: Cloud Institute.

Conn, S. A. (1995). When the earth hurts, who responds? In T. Roszak, M. E. Gomes, & A. D. Kanner (Eds.), *Ecopsychology: Restoring the earth, healing the mind* (pp. 156-171). San Francisco, CA: Sierra Club Books.

Connect2Compete. (2013). Connect2Compete launches nationwide campaign to increase digital literacy, internet access for 100 million "offline" Americans [Press release]. Retrieved from http://www.prnewswire.com/news-releases/connect2compete-launches-nationwide-campaign-to-increase-digital-literacy-internet-access-for-100-million-offline-americans-199322141.html

Cook, S. D. N. (2006). Technological revolutions and the Gutenberg myth. In R. Hassan, & J. Thomas (Eds.), *The new media theory reader* (pp. 11-18). New York: Open University Press.

Corbett, J. B. (2006). *Communicating nature: How we create and understand environmental messages.* Washington, DC: Island Press.

Cortés, C. (2005). How the media teach. In G. Schwarz, P. U. Brown, & National Society for the Study of Education (Eds.), *Media literacy: Transforming curriculum and teaching* (pp. 55-73). Malden, MA: Wiley-Blackwell.

Coupe, L. (2000a). General introduction. In L. Coupe (Ed.), *The green studies reader: From romanticism to ecocriticism* (pp. 1-8). London, U.K.: Routledge.

Coupe, L. (Ed.). (2000b). *The green studies reader: From romanticism to ecocriticism*. London, U.K.: Routledge.

Coupland, N., & Jaworski, A. (Eds.). (2006a). *The discourse reader*. London: Routledge.

Coupland, N., & Jaworski, A. (2006b). Introduction: Perspectives on discourse analysis. In N. Coupland, & A. Jaworski (Eds.), *The discourse reader*. London: Routledge.

Cox, R. (2009). *Environmental communication and the public sphere* (2nd ed.). Thousand Oaks, CA: Sage.

Coyle, K. (2005). *Environmental literacy in America*. Washington, DC: National Environmental Education & Training Foundation.

Cross, J. (2007). *Informal learning: Rediscovering the natural pathways that inspire innovation and performance*. San Francisco, CA: Pfeiffer/Wiley.

Cubitt, S., Hassan, R., & Volkmer, I. (2011). Does cloud computing have a silver lining? *Media, Culture & Society, 33*(1), 149-158. doi:10.1177/0163443710382974

Curry, P. (2006). *Ecological ethics: An introduction*. Malden, MA: Polity Press.

De Abreu, B. S. (2011). *Media literacy, social networking, and the web 2.0 environment for the K-12 educator*. New York: Peter Lang.

DeFleur, M. L., & Dennis, E. E. (2002). *Understanding mass communication: A liberal arts perspective*. Boston, MA: Houghton Mifflin.

Deloria, V., & Wildcat, D. (2001). *Power and place: Indian education in America*. Golden, CO: Fulcrum.

DeLuca, K. M. (1999). *Image politics: The new rhetoric of environmental activism*. New York: Guilford Press.

Domaille, K. (2012). The professional preparation, progression and development of media teachers. In E. Scarratt, & J. Davison (Eds.), *The media teacher's handbook* (pp. 225-239). New York: Routledge.

Dominick, J. R. (2009). *The dynamics of mass communication: Media in the digital age* (10th ed.). New York: McGraw-Hill Higher Education.

Downes, S. (2007). Models for sustainable open educational resources. *Interdisciplinary Journal of Knowledge and Learning Objects, 3*, 30-44.

Dryzek, J. S. (2005). *The politics of the earth: Environmental discourses*. Oxford, U.K.: Oxford University Press.

Du Gay, P., Hall, S., Janes, L., & Mackay, H. (1997). *Doing cultural studies: The story of the Sony Walkman*. London: Sage, in association with The Open University.

Editors of Rethinking Schools. (2011). Our climate crisis is an education crisis. Retrieved from http://www.rethinkingschools.org/archive/25_03/edit253.shtml

Edwards, A. R. (2005). *The sustainability revolution: Portrait of a paradigm shift*. Gabriola, B.C.: New Society Publishers.

Edwards, A. R. (2010). *Thriving beyond sustainability: Pathways to a resilient society*. Gabriola Island, B.C.: New Society Publishers.

Ehrenfeld, J. (2008). *Sustainability by design: A subversive strategy for transforming our consumer culture*. New Haven, CT: Yale University Press.

Ellul, J. (1964). *The technological society* [Technique.] (1st American ed.). New York: Knopf.

Esbjorn-Hargens, S., & Zimmerman, M. E. (2009). *Integral ecology: Uniting multiple perspectives on the natural world*. Boston, MA: Integral Books.

Esteva, G., & Prakash, M. S. (1998). *Grassroots post-modernism: Remaking the soil of cultures*. London: Zed Books.

Evans, T. L. (2012). *Occupy education: Learning and living sustainability*. New York: Peter Lang.

Ewen, S. (2001). *Captains of consciousness: Advertising and the social roots of the consumer culture* (25th anniversary ed.). New York: Basic Books.

Fairclough, N. (1999). Global capitalism and critical awareness of language. *Language Awareness*, 8(2), 71-83. doi:10.1080/09658419908667119

Fairclough, N., & Wodak, R. (1997). Critical discourse analysis. In T. A. Van Dijk (Ed.), *Discourse studies: A multidisciplinary introduction* (pp. 258-284). Thousand Oaks, CA: Sage.

Fisherkeller, J. (2011). *International perspectives on youth media: Cultures of production and education*. New York: Peter Lang.

Foucault, M. (1998). Different spaces. In J. D. Faubion (Ed.), *Aesthetics, method, and epistemology* (pp. 175-186). New York: New Press.

Fuller, M. A. (2005). *Media ecologies: Materialist energies in art and technoculture*. Cambridge, MA: MIT Press.

Gadotti, M. (2010). Pedagogy of the earth and the culture of sustainability. *Journal of Education for Sustainable Development*, 4(2), 203-211. doi:10.1177/097340821000400207

Garrard, G. (2004). *Ecocriticism*. London: Routledge.

Garrard, G. (2009). Ecocriticism: The ability to investigate cultural artifacts from an ecological perspective. In A. Stibbe (Ed.), *The handbook of sustainability literacy: Skills for a changing world*. Totnes, U.K.: Green Books.

Gauntlett, D. (2007). Media studies 2.0. Retrieved from http://www.theory.org.uk/mediastudies2.htm

Gauntlett, D. (2011). *Making is connecting: The social meaning of creativity from DIY and knitting to YouTube and web 2.0*. Cambridge, U.K.: Polity Press.

Gee, J. P. (2011a). *How to do discourse analysis: A toolkit*. New York: Routledge.

Gee, J. P. (2011b). *An introduction to discourse analysis: Theory and method*. Milton Park, Abingdon, U.K.: Routledge.

Gee, J. P., & Hayes, E. (2011). *Language in a digital age*. New York: Routledge.

Gerbner, G. (1998). Introduction: Why the cultural environment movement? *Gazette: International Journal for Communication Studies*, 60(2), 133.

Giddens, A. (1984). *The constitution of society: Outline of the theory of structuration.* Berkeley: University of California Press.

Gieryn, T. F. (1983). Boundary-work and the demarcation of science from non-science: Strains and interests in professional ideologies of scientists. *American Sociological Review, 48*(6), 781-795.

Gilding, P. (2012). *The great disruption: Why the climate crisis will bring on the end of shopping and the birth of a new world.* New York: Bloomsbury Press.

Giroux, H. A. (1994). *Disturbing pleasures: Learning popular culture.* New York: Routledge.

Glendinning, C. (1994). *My name is Chellis and I'm in recovery from western civilization.* Boston, MA: Shambhala.

Glendinning, C. (1995). Recovery from western civilization. In G. Sessions (Ed.), *Deep ecology for the twenty-first century* (pp. 37-40). Boston, MA: Shambhala.

Goleman, D., Bennett, L., & Barlow, Z. (2012). *Eco literate: How educators are cultivating emotional, social, and ecological intelligence.* San Francisco, CA: Jossey-Bass.

Golley, F. B. (1998). *A primer for environmental literacy.* New Haven, CT: Yale University Press.

Gramsci, A., Hoare, Q., & Smith, G. N. (2005). *Selections from the prison notebooks of Antonio Gramsci.* New York: International.

Gray, J. (2006). *Watching with the Simpsons: Television, parody, and intertextuality.* New York: Routledge.

Green, N. (2006). On the move: Technology, mobility, and the mediation of social time and space. In R. Hassan & J. Thomas (Eds.), *The new media theory reader* (pp. 244-248). Maidenhead, U.K.: Open University Press.

Greenpeace International. (2010, March 30). *Make IT green: Cloud computing and its contribution to climate change.* Greenpeace Report. Retrieved from http://www.greenpeace.org/usa/press-center/reports4/make-it-green-cloud-computing

Greenwood, D. J., & Levin, M. (2007). *Introduction to action research: Social research for social change* (2nd ed.). Thousand Oaks, CA: Sage.

Grigorov, S. K., & Matias Fleuri, R. (2012). Ecopedagogy: Educating for a new eco-social intercultural perspective. *Visão Global, 15*(1-2), 433-454.

Guattari, F. (2008). *Three ecologies.* London: Continuum.

Gurell, S., Kuo, Y., & Walker, A. (2010). The pedagogical enhancement of open education: An examination of problem-based learning. *International Review of Research in Open and Distance Learning, 11*(2). Retrieved from http://www.irrodl.org/index.php/irrodl/article/view/886/1633

Gutiérrez-Martín, A., & Tyner, K. (2012). Educación para los medios, alfabetización mediática y competencia digital. *Revista Comunicar, XIX*(38), 31-39.

Hansen, A. (2009). *Environment, media and communication.* New York: Routledge.

Harding, S. (2006). *Animate earth: Science, intuition and Gaia.* White River Junction, VT: Chelsea Green.

Hardt, M., & Negri, A. (2004). *Multitude: War and democracy in the age of empire*. New York: Penguin Press.
Harlen, W., & Doubler, S. J. (2007). Researching the impact of online professional development for teachers. In R. Andrews & C. Haythornthwaite (Eds.), *The Sage handbook of e-learning research* (pp. 466-486). London: Sage. doi:10.4135/9781848607859.n21
Hartley, J., Montgomery, M., Rennie, E., & Brennan, M. (2002). *Communication, cultural and media studies: The key concepts* (3rd ed.). London: Routledge.
Hartley, J. (2012). Digital futures for cultural and media studies. Malden, MA: Wiley-Blackwell.
Harvey, D. (2005). *A brief history of neoliberalism*. Oxford, U.K.: Oxford University Press.
Harvey, D. (2008, September/October). The right to the city. *New Left Review, 53*.
Hassan, R., & Thomas, J. (Eds.). (2006). *The new media theory reader*. New York: Open University Press.
Hawken, P. (2007). *Blessed unrest: How the largest movement in the world came into being, and why no one saw it coming*. New York: Viking.
Heise, U. K. (2002). Unnatural ecologies: The metaphor of the environment in media theory. *Configurations, 10*(1), 149-168.
Heise, U. K. (2008). *Sense of place and sense of planet: The environmental imagination of the global*. Oxford, U.K.: Oxford University Press.
Henrich, J., Heine, S. J., & Norenzayan, A. (2010). The weirdest people in the world. *Behavioral and Brain Sciences, 33*(2-3), 61-83.
Herman, E. S., & Chomsky, N. (2002). *Manufacturing consent: The political economy of the mass media*. New York: Pantheon Books.
Heron, J., & Reason, P. (2006). The practice of co-operative inquiry: Research 'with' rather than 'on' people. In P. Reason, & H. Bradbury (Eds.), *Handbook of action research: The concise paperback edition* (pp. 144-154). Thousand Oaks, CA: Sage.
Hobbs, R. (2011). *Digital and media literacy: Connecting culture and classroom*. Thousand Oaks, CA: Corwin Press.
Hoff, B. (1983). *The Tao of pooh*. New York: Penguin Books.
Holmgren, D. (2002). *Permaculture: Principles and pathways beyond sustainability*. Hepburn, Victoria, Australia: Holmgren Design Services.
Hornborg, A. (2001). *The power of the machine: Global inequalities of economy, technology, and environment*. Walnut Creek, CA: AltaMira Press.
Hughes, T. (2005). *Human-built world: How to think about technology and culture*. Chicago: University of Chicago Press.
Hustwit, G. (Director). (2009). Objectified [Motion picture]. United States: Plexi Productions.
Illich, I. (1971). *Deschooling society*. New York: Harper & Row.
Illich, I. (1973). *Tools for conviviality*. London, U.K.: Calder & Boyars.
Innis, H. A. (1999). *The bias of communication*. Toronto, Canada: University of Toronto Press.

Invisible Children. (2012). *Kony 2012*. Retrieved from https://www.youtube.com/watch?v=Y4MnpzG5Sqc

Jacobs, H. H. (2010). A new essential curriculum for a new time. In H. H. Jacobs (Ed.), *Curriculum 21: Essential education for a changing world* (pp. 168-185). Alexandria, VA: Association for Supervision and Curriculum Development.

Jacobs, J. (2001). *The nature of economies*. New York: Vintage.

Jagtenberg, T., & McKie, D. (1997). *Eco-impacts and the greening of postmodernity: New maps for communication studies, cultural studies, and sociology*. Thousand Oaks, CA: Sage.

Jenkins, H. (2006). *Convergence culture: Where old and new media collide*. New York: New York University Press.

Jenkins, H., Puroshotma, R., Clinton, K., Weigel, M., & Robison, A. J. (2005). *Confronting the challenges of participatory culture: Media education for the 21st century*. MacArthur Foundation. Cambridge, MA: MIT Press. Retrieved from http://www.macfound.org/press/publications/white-paper-confronting-the-challenges-of-participatory-culture-media-education-for-the-21st-century-by-henry-jenkins/

Jensen, K. B. (2002a). Introduction. In K. B. Jensen (Ed.), *A handbook of media and communications research: Qualitative and quantitative methodologies* (pp. 1-11). London, U.K.: Routledge.

Jensen, K. B. (2002b). The qualitative research process. In K. B. Jensen (Ed.), *A handbook of media and communications research: Qualitative and quantitative methodologies* (pp. 235-253). London, U.K.: Routledge.

Kagan, S. (2011). *Art and sustainability: Connecting patterns for a culture of complexity*. New Brunswick, NJ: Transcript.

Kahn, R. (2010). *Critical pedagogy, ecoliteracy, and planetary crisis: The ecopedagogy movement*. New York: Peter Lang.

Kahn, R. (2011). Technoliteracy at the sustainability crossroads: Posing ecopedagogical problems for digital literacy frameworks. In P. Trifonas (Ed.), *Learning the virtual life: Public pedagogy in digital world* (pp. 43-62). New York: Routledge.

Kamenetz, A. (2010). *DIY U: Edupunks, edupreneurs, and the coming transformation of higher education*. White River Junction, VT: Chelsea Green.

Kanner, A. D., & Gomes, M. E. (1995). The all-consuming self. In T. Roszak, M. E. Gomes, & A. D. Kanner (Eds.), *Ecopsychology: Restoring the earth, healing the mind* (pp. 77-91). San Francisco, CA: Sierra Club Books.

Kellner, D. (2003). *Media spectacle*. New York: Routledge.

Kellner, D., & Share, J. (2007). Critical media literacy, democracy, and the reconstruction of education. In D. P. Macedo & S. R. Steinberg (Eds.), *Media literacy: A reader* (pp. 3-23). New York: Peter Lang.

Kelly, K. (2008). Better than free. *ChangeThis*, issue 53-01. Retrieved from http://changethis.com/manifesto/show/53.01.BeyondFree

Kendall, A., & McDougall, J. (2012). Critical media literacy after the media. *Revista Comunicar, XIX*(38), 21–29.

Kenner, R., & Pearlstein, E. (Producers), & Kenner, R. (Director). (2008). *Food, Inc.* [Motion picture]. United States: Magnolia Pictures.

Korten, D. C. (2006). *The great turning: From empire to earth community.* San Francisco, CA: Berrett-Koehler.

Kuhn, T. S. (1996). *The structure of scientific revolutions* (3rd ed.). Chicago: University of Chicago Press.

LaChapelle, D. (1995). Ritual: The pattern that connects. In G. Sessions (Ed.), *Deep ecology for the twenty-first century* (pp. 57–63). Boston, MA: Shambhala.

Lakoff, G., & Johnson, M. (1980). *Metaphors we live by.* Chicago, IL: University of Chicago Press.

Lanier, J. (2010). *You are not a gadget: A manifesto.* New York: Alfred A. Knopf.

Lappé, F. M. (2011). *EcoMind: Changing the way we think, to create the world we want.* New York: Nation Books.

Lasn, K. (2000). *Culture jam: How to reverse America's suicidal consumer binge–and why we must.* New York: Quill.

Laughey, D. (2007). *Key themes in media theory.* New York: Open University Press.

Lave, J., & Wenger, E. (1991). *Situated learning: Legitimate peripheral participation.* New York: Cambridge University Press.

Leadbeater, C., & Powell, D. (2009). *We-think: Mass innovation, not mass production.* London, U.K.: Profile.

Leiss, W. (1972). *The domination of nature.* Boston, MA: Beacon Press.

Leonard, A. (2007). *Story of stuff, referenced and annotated script.* Retrieved from http://act.storyofstuff.org/page/-/Downloads/MovieScripts/SoStuff_Annotated_Script.pdf

Leopold, A. (1987). *A Sand County almanac, and sketches here and there.* New York: Oxford University Press.

Lessig, L. (2008). *Remix: Making art and commerce thrive in the hybrid economy.* New York: Penguin Press.

Lévy, P. (1998). *Becoming virtual: Reality in the digital age.* New York: Plenum Trade.

Lewis, J., & Boyce, T. (2009). Climate change and the media: The scale of the challenge. In J. Lewis, & T. Boyce (Eds.), *Climate change and the media (global crises and the media)* (pp. 3–16). New York: Peter Lang.

Lewis, J., & Jhally, S. (1998). The struggle over media literacy. *Journal of Communication, 48*(1), 109.

Lippmann, W. (2007). *Public opinion.* Miami, FL: BN.

Liska, J. R., & Cronkhite, G. (1995). *An ecological perspective on human communication theory.* Fort Worth, TX: Harcourt Brace College.

Logan, R. K. (2007). The biological foundation of media ecology. *Explorations Media Ecology*, 6(1), 19–34.

López, A. (2008). *Mediacology: A multicultural approach to media literacy in the 21st century*. New York: Peter Lang.

López, A. (2010). Defusing the cannon/canon: An organic media approach to environmental communication. *Environmental Communication: A Journal of Nature and Culture*, 4(1), 99–108.

López, A. (2011a). Greening a digital media course. Retrieved from http://www.newmedialiteracies.org/2011/05/greening_a_digital_media_cours/

López, A. (2011b). Greening media education. Retrieved from http://www.manifestoformediaeducation.co.uk/2011/02/antonio-lopez/

López, A. (2011c). Practicing sustainable youth media. In J. Fisherkeller (Ed.), *International perspectives on youth media: Cultures of production and education* (pp. 317–337). New York: Peter Lang.

López, A. (2012). *The media ecosystem: What ecology can teach us about responsible media practice*. Berkeley, CA: North Atlantic Books.

López, A. (2013, Winter). Greening a digital media course: A field report. *Journal for Sustainability Education*.

Luhmann, N. (1989). *Ecological communication*. Chicago, IL: University of Chicago Press.

Lum, C. M. K. (2006). Notes toward an intellectual history of media ecology. In C. M. K. Lum (Ed.), *Perspectives on culture, technology and communication: The media ecology*. Cresskill, NJ: Hampton Press.

Lusted, D. (Ed.). (1991). *The media studies book: A guide for teachers*. London: Routledge.

Machin, D., & Mayr, A. (2012). *How to do critical discourse analysis: A multimodal approach*. Los Angeles, CA: Sage.

Madden, M., Lenhart, A., Duggan, M., Cortesi, S. & Gasser, U. (2013). Teens and technology 2013. *Pew Research Internet Project*. Retrieved from http://www.pewinternet.org/Reports/2013/Teens-and-Tech.aspx

Malick, T. (Director). (2005). *The new world* [Motion picture]. United States: New Line Cinema.

Mander, J. (1991). *In the absence of the sacred: The failure of technology and the survival of the Indian nations*. San Francisco, CA: Sierra Club Books.

Mander, J. (1995). Leaving the earth: Space colonies, Disney, and EPCOT. In G. Sessions (Ed.), *Deep ecology for the twenty-first century* (pp. 311–319). Boston, MA: Shambhala.

Mander, J. (2002). *Four arguments for the elimination of television*. New York: Perennial.

Mander, J., & Goldsmith, E. (Eds.). (1996). *The case against the global economy: And for a turn toward the local*. San Francisco, CA: Sierra Club Books.

Maser, C. (1999). *Vision and leadership in sustainable development*. Boca Raton, FL: Lewis.

Mason, P. (2012). *Why it's kicking off everywhere: The new global revolutions*. London, U.K.: Verso.

Masterman, L. (1989). *Teaching the media*. London, U.K.: Routledge.

Maturana, H. R., & Varela, F. J. (1998). *The tree of knowledge: The biological roots of human understanding*. Boston, MA: Shambhala.

Maxwell, R., & Miller, T. (2009). Talking rubbish: Green citizenship, media and the environment. In J. Lewis, & T. Boyce (Eds.), *Climate change and the media (global crises and the media)* (pp. 17–27). New York: Peter Lang.

Maxwell, R., & Miller, T. (2012). *Greening the media*. New York: Oxford University Press.

Mayes, T., & de Freitas, S. (2007). Learning and e-learning: The role of theory. In H. Beetham & R. Sharpe (Eds.), *Rethinking pedagogy for a digital age: Designing and delivering e-learning*. London, U.K.: Routledge.

McChesney, R. W. (1999). *Rich media, poor democracy: Communication politics in dubious times*. Urbana: University of Illinois Press.

McCullough, M. (2004). *Digital ground: Architecture, pervasive computing, and environmental knowing*. Cambridge, MA: MIT Press.

McDougall, J., & Potamitis, N. (2010). *The media teacher's book*. London, U.K.: Hodder Education.

McGilchrist, I. (2009). *The master and his emissary: The divided brain and the making of the western world*. New Haven, CT: Yale University Press.

McKibben, B. (2008). *Deep economy: The wealth of communities and the durable future*. New York: Holt Paperbacks.

McLuhan, M. (2002a). *The Gutenberg galaxy: The making of typographic man*. Toronto, Canada: University of Toronto Press.

McLuhan, M. (2002b). *Understanding media: The extensions of man*. Cambridge, MA: MIT Press.

McLuhan, M., & Powers, B. R. (1989). *The global village: Transformations in world life and media in the 21st century*. New York: Oxford University Press.

Meadows, D. H. (1991). *The global citizen*. Washington, DC: Island Press.

Meadows, D. H. (2009). *Thinking in systems: A primer*. London: Earthscan.

Meadows, D. H., Randers, J., & Meadows, D. L. (2004). *The limits to growth: The 30-year update*. White River Junction, VT: Chelsea Green.

Media & Learning. (n.d.). *Media education and literacy: Equipping learners for open, creative learning futures*. Retrieved from http://www.media-and-learning.eu/

Merchant, C. (1989). *The death of nature: Women, ecology, and the scientific revolution*. New York: HarperOne.

Merchant, C. (2008). *Key concepts in social theory: Ecology (Second edition)*. Atlantic Highlands, NJ: Humanity Books.

Meyrowitz, J. (1980). *Analyzing media: Metaphors as methodologies*. Paper presented at the New England Conference on Teaching Students to Think, Amherst, MA: Retrieved from http://www.eric.ed.gov/ERICWebPortal/contentdelivery/servlet/ERICServlet?accno=ED206030

Meyrowitz, J. (1985). *No sense of place: The impact of electronic media on social behavior.* New York: Oxford University Press.
Meyrowitz, J. (1998, Winter). Multiple media literacies. *Journal of Communication,* 96-108.
Milstein, T. (2009). Environmental communication theories. In S. Littlejohn, & K. Foss (Eds.), *Encyclopedia of communication theory* (pp. 344-349). Thousand Oaks, CA: Sage.
Milstein, T., & Dickinson, E. (2012). Gynocentric greenwashing: The discursive gendering of nature. *Communication, Culture & Critique,* 5(4), 510-532. doi:10.1111/j.1753-9137.2012.01144.x
Moores, S. (2012). *Media, place and mobility.* New York: Palgrave Macmillan.
Morley, D. (2005a). Communication. In T. Bennett, L. Grossberg, & M. Morris (Eds.), *New keywords: A revised vocabulary of culture and society* (pp. 47-50). Malden, MA: Blackwell.
Morley, D. (2005b). Media. In T. Bennett, L. Grossberg, & M. Morris (Eds.), *New keywords: A revised vocabulary of culture and society* (pp. 212-214). Malden, MA: Blackwell.
Morozov, E. (2011). *The net delusion: The dark side of internet freedom.* New York: PublicAffairs.
Morris, D., & Martin, S. (2009). Complexity, systems thinking and practice. In A. Stibbe (Ed.), *The handbook of sustainability literacy: Skills for a changing world.* Totnes, U.K.: Green Books.
Morrow, R. A., & Brown, D. D. (1994). *Critical theory and methodology.* Thousand Oaks, CA: Sage.
Mulvey, L. (2001). Visual pleasure and narrative cinema. In M. G. Durham, & D. Kellner (Eds.), *Media and cultural studies: Keyworks* (Rev. ed., pp. 393-404). Malden, MA: Blackwell.
Mumford, L. (1967). *The myth of the machine: Technics and human development.* New York: Harcourt Brace Jovanovich.
Mumford, L. (1970). *The pentagon of power.* New York: Harcourt Brace Jovanovich.
Naess, A. (1995). The deep ecology movement. In G. Sessions (Ed.), *Deep ecology for the twenty-first century* (pp. 64-84). Boston, MA: Shambhala.
Nardi, B. A., & O'Day, V. (2000). *Information ecologies: Using technology with heart.* Cambridge, MA: MIT Press.
National Association for Media Literacy Education. (2007, November). Core principles of media literacy education in the United States. Retrieved from http://namle.net/wp-content/uploads/2013/01/CorePrinciples.pdf
Naughton, J. J. (2006). Blogging and the emerging media ecosystem. Retrieved from https://reutersinstitute.politics.ox.ac.uk/fileadmin/documents/discussion/blogging.pdf
Naughton, J. J. (2012). *From Gutenberg to Zuckerberg: What you really need to know about the internet.* London: Quercus.
Nelson, A. (1998). *The learning wheel: Ideas and activities for multicultural and holistic lesson planning.* Evergreen, CO: WHEEL Council.

Neuzil, M., & Kovarik, W. (1996). *Mass media and environmental conflict: America's green crusades.* Thousand Oaks, CA: Sage.

Nisbett, R. E. (2004). *The geography of thought: How Asians and Westerners think differently–and why.* New York: Free Press.

Nye, D. E. (1994). *American technological sublime.* Cambridge, MA: MIT Press.

Nye, D. E. (2006). The consumer's sublime. In R. Hassan, & J. Thomas (Eds.), *The new media theory reader* (pp. 27–38). New York: Open University Press.

O'Connor, J., & McDermott, I. (1997). *The art of systems thinking: Essential skills for creativity and problem solving.* London: Thorsons.

Odum, E. P., & Barrett, G. W. (2005). *Fundamentals of ecology* (5th ed.). Belmont, CA: Thomson Brooks/Cole.

Ong, W. J. (1982). *Orality and literacy: The technologizing of the word.* London: Methuen.

Orr, D. W. (1994). *Earth in mind: On education, environment, and the human prospect.* Washington, DC: Island Press.

O'Sullivan, E., & Taylor, M. M. (2004). Introduction: Conundrum, challenge, and choice. In E. O'Sullivan, & M. M. Taylor (Eds.), *Learning toward an ecological consciousness: Selected transformative practices* (pp. 1–4). New York: Palgrave Macmillan.

Parenti, M. (1986). *Inventing reality: The politics of the mass media.* New York: St. Martin's Press.

Parr, A. (2009). *Hijacking sustainability.* Cambridge, MA: MIT Press.

Pendleton-Jullian, A. (2009). *Design education and innovative ecotones* Retrieved from https://fourplusone.wordpress.com/design-education-and-innovation-ecotones/

Peters, J. D. (1999). *Speaking into the air: A history of the idea of communication.* Chicago, IL: University of Chicago Press.

Pink, D. H. (2005). *A whole new mind: Moving from the information age to the conceptual age.* New York: Riverhead Books.

Pittman, J. (2004). Living sustainably through higher education: A whole systems design approach to organizational change. In P. B. Corcoran & A. E. J. Wals (Eds.), *Higher education and the challenge of sustainability: Problematics, promise, and practice* (pp. 199–212). Boston, MA: Kluwer Academic.

Pollan, M. (2006). *The omnivore's dilemma: A natural history of four meals.* New York: Penguin Press.

Postman, N. (1993). *Technopoly: The surrender of culture to technology.* New York: Vintage Books.

Postman, N. (1998, March 28). Five things we need to know about technological change. Retrieved from http://www.cs.ucdavis.edu/~rogaway/classes/188/materials/postman.pdf

Potter, W. J. (2004). *Theory of media literacy: A cognitive approach.* Thousand Oaks, CA: Sage.

Rasmussen, D. (2000). Our life out of balance: The rise of literacy and the demise of pattern languages. *Encounter: Education for Meaning and Social Justice, 1*(13), 13–21.

Rayner, P., Wall, P., & Kruger, S. (2004). *Media studies: The essential resource*. Abingdon, Oxon, U.K.: Routledge.

Reason, P., & Bradbury, H. (2006). Introduction: Inquiry and participation in search of a world worthy of human aspiration. In P. Reason & H. Bradbury (Eds.), *Handbook of action research: The concise paperback edition* (pp. 1-14). Thousand Oaks, CA: Sage.

Reddy, M. J. (1979). The conduit metaphor: A case of frame conflict in our language about language. In A. Ortony (Ed.), *Metaphor and thought* (pp. 284-324). Cambridge, U.K.: Cambridge University Press.

Reid, G., Kennedy, S., & Ferstad, C. (Producers), & Reid, G. (Director). (2013). *Overview* [Motion picture]. United States: Planetary Collective.

Rheingold, H. (2002). *Smart mobs: The next social revolution*. Cambridge, MA: Perseus.

Rheingold, H. (2012). *Net smart: How to thrive online*. Cambridge, MA: MIT Press.

Rifkin, J. (2011). *The third industrial revolution: How lateral power is transforming energy, the economy, and the world*. New York: Palgrave Macmillan.

Romanyshyn, R. D. (1989). *Technology as symptom and dream*. London, U.K.: Routledge.

Roszak, T. (1995). Where psyche meets gaia. In T. Roszak, M. E. Gomes, & A. D. Kanner (Eds.), *Ecopsychology: Restoring the earth, healing the mind* (pp. 1-17). San Francisco, CA: Sierra Club Books.

Rowe, J. (2008). The parallel economy of the commons. In T. W. Institute (Ed.), *State of the world 2008: Toward a sustainable global economy* (pp. 138-150). New York: W.W. Norton.

RSA. (2010). RSA animate: Changing education paradigms. Retrieved from https://www.youtube.com/watch?v=zDZFcDGpL4U

Rushkoff, D. (2011). *Program or be programmed: Ten commands for a digital age*. Berkeley, CA: Soft Skull.

Sachs, J. (2012). *Winning the story wars: Why those who tell—and live—the best stories will rule the future*. Boston, MA: Harvard Business Review Press.

Sale, K. (1996). *Rebels against the future: The Luddites and their war on the Industrial Revolution. Lessons for the computer age*. Reading, MA: Addison-Wesley.

Samovar, L. A., Porter, R. E., McDaniel, E. R., & Roy, C. S. (2012). *Communication between cultures*. Boston, MA: Wadsworth Cengage Learning.

Scannell, P. (2002). History, media and communication. In K. B. Jensen (Ed.), *A handbook of media and communications research: Qualitative and quantitative methodologies* (pp. 191-205). London: Routledge.

Scarratt, E., & Davison, J. (Eds.). (2012). *The media teacher's handbook*. Abingdon, Oxon, U.K.: Routledge.

Scheibe, C., & Rogow, F. (2012). *The teacher's guide to media literacy: Critical thinking in a multimedia world*. Thousand Oaks, CA: Corwin.

Schumacher, E. F. (1993). *Small is beautiful: A study of economics as if people mattered.* London: Vintage.

Senge, P. M. (2008). *The necessary revolution: How individuals and organizations are working together to create a sustainable world.* New York: Doubleday.

Senge, P. M., Scharmer, C. O., Jaworski, J., & Flowers, B. S. (2005). *Presence: Exploring profound change in people, organizations, and society.* New York: Doubleday.

Sewall, L. (1995). The skill of ecological perception. In T. Roszak, M. E. Gomes & A. D. Kanner (Eds.), (pp. 201-215). San Francisco, CA: Sierra Club Books.

Share, J. (2009). *Media literacy is elementary: Teaching youth to critically read and create media.* New York: Peter Lang.

Shepard, P. (1995). Nature and madness. In T. Roszak, M. E. Gomes, & A. D. Kanner (Eds.), *Ecopsychology: Restoring the earth, healing the mind* (pp. 21-40). San Francisco, CA: Sierra Club Books.

Shepard, P. (1998). *Nature and madness.* Athens: University of Georgia Press.

Sheridan, M. P., & Rowsell, J. (2010). *Design literacies: Learning and innovation in the digital age.* London: Routledge.

Shirky, C. (2008). *Here comes everybody: How digital networks transform our ability to gather and cooperate.* New York: Penguin Press.

Shirky, C. (2010). *Cognitive surplus: Creativity and generosity in a connected age.* New York: Penguin Press.

Shiva, V. (1993). *Monocultures of the mind: Perspectives on biodiversity and biotechnology.* London: Zed Books.

Shiva, V. (2005). *Earth democracy: Justice, sustainability, and peace.* Cambridge, MA: South End Press.

Shiva, V. (2008). *Soil not oil: Environmental justice in a time of climate crisis.* Cambridge, MA: South End Press.

Shlain, L. (1998). *The alphabet versus the goddess: The conflict between word and image.* New York: Viking.

Silver, J. (Producer), Wachowski, A. & Wachowski, L. (Directors). (1999). *The matrix* [Motion picture]. United States: Warner Bros. Pictures.

Silverstone, R. (2007). *Media and morality: On the rise of the mediapolis.* Cambridge, U.K.: Polity Press.

Slack, J. D. (2005). Environment/ecology. In T. Bennett, L. Grossberg, & M. Morris (Eds.), *New keywords: A revised vocabulary of culture and society* (pp. 106-109). Malden, MA: Blackwell.

Snyder, G. (1995). Cultured or crabbed. In G. Sessions (Ed.), *Deep ecology for the twenty-first century* (pp. 47-49). Boston, MA: Shambhala.

Sontag, S. (2002, December 9). Looking at war. *The New Yorker*, 82-98.

Sperry, S. (2011). *Media constructions of sustainability: Food, water and agriculture*. Ithaca, NY: Project Look Sharp.

Star, S. L., & Griesemer, J. R. (1989). Institutional ecology, 'translations' and boundary objects: Amateurs and professionals in Berkeley's museum of vertebrate zoology, 1907-1939. *Social Studies of Science, 19*(3), 387-420.

Steele, J. R., & Brown, J. D. (1995). Adolescent room culture: Studying media in the context of everyday life. *Journal of Youth and Adolescence, 24*(5), 551-576.

Sterling, S. (2004). *Sustainable education: Re-visioning learning and change*. Devon, U.K.: Green Books.

Sterling, S. (2009). Ecological intelligence: Viewing the world relationally. In A. Stibbe (Ed.), *The handbook of sustainability literacy: Skills for a changing world*. Totnes, U.K.: Green Books.

Stibbe, A. (2009). *The handbook of sustainability literacy: Skills for a changing world*. Totnes, U.K.: Green Books.

Stibbe, A., & Luna, H. (2009). Introduction. In A. Stibbe (Ed.), *The handbook of sustainability literacy: Skills for a changing world*. Totnes, U.K.: Green Books.

Story of Stuff Project. (2009). Story of stuff. Retrieved from http://www.storyofstuff.org/movies-all/story-of-stuff/

Sturken, M., & Cartwright, L. (2009). *Practices of looking: An introduction to visual culture*. New York: Oxford University Press.

Sunstein, C. (2006). Citizens. In R. Hassan, & J. Thomas (Eds.), *The new media theory reader* (pp. 203-211). New York: Open University Press.

Swisher, K., & Deyhle, D. (1992). Adapting instruction to culture. In J. A. Reyhner (Ed.), *Teaching American Indian students*. Norman: University of Oklahoma Press.

Thomas, D., & Brown, J. S. (2011). *A new culture of learning: Cultivating the imagination for a world of constant change*. Lexington, KY: CreateSpace.

Thomas, S. (1995). Myths in and about television. In J. Downing, A. Mohammadi, & A. Sreberny (Eds.), *Questioning the media: A critical introduction* (pp. 444-459). Thousand Oaks, CA: Sage.

Thomashow, M. (1995). *Ecological identity: Becoming a reflective environmentalist*. Cambridge, MA: MIT Press.

Thomashow, M. (2003). *Bringing the biosphere home: Learning to perceive global environmental change*. Cambridge, MA: MIT Press.

Todd, N. J., & Todd, J. (1994). *From eco-cities to living machines: Principles of ecological design*. Berkeley, CA: North Atlantic Books.

Todd, Z., & Harrison, S. J. (2010). Metaphor analysis. In S. N. Hesse-Biber & P. Leavy (Eds.), *Handbook of emergent methods* (pp. 479-493). New York: Guilford Press.

Tomlinson, B. (2010). *Greening through IT: Information technology for environmental sustainability*. Cambridge, MA: MIT Press.

Tracy, C. M. (2012). *The newsphere: Understanding the news and information environment*. New York: Peter Lang.

Traina, F. (1995). The challenge of bioregional education. In F. Traina, & S. Darley-Hill (Eds.), *Perspectives in bioregional education* (pp. 19–26). Troy, MI: North American Association for Environmental Education.

Turkle, S. (2011). *Alone together: Why we expect more from technology and less from each other*. New York: Basic Books.

Tyner, K. (1991). The media education elephant. Retrieved from http://www.medialit.org/reading-room/media-education-elephant

Tyner, K. (1998). *Literacy in a digital world: Teaching and learning in the age of information*. Mahwah, NJ: Erlbaum.

Tyner, K. (2010). Introduction: New agendas for media literacy. In K. Tyner (Ed.), *Media literacy: New agendas in communication* (New Agendas in Communication series) (pp. 1–7). New York: Routledge.

Tyner, K. (2011). New agendas for media literacy. Retrieved from http://www.manifestoformediaeducation.co.uk/category/kathleen-tyner/

UNESCO. (2012). Education for sustainable development (ESD). Retrieved from http://www.unesco.org/new/en/education/themes/leading-the-international-agenda/education-for-sustainable-development/

Varela, F. J. (1999). *Ethical know-how: Action, wisdom, and cognition*. Stanford, CA: Stanford University Press.

Varela, F. J., Thompson, E., & Rosch, E. (1991). *The embodied mind: Cognitive science and human experience*. Cambridge, MA: MIT Press.

Walljasper, J. (2010). *All that we share: How to save the economy, the environment, the internet, democracy, our communities, and everything else that belongs to all of us*. New York: New Press.

Warshall, P. (2012). Introduction. In P. Warshall, K. Ausubel, A. Mangan, & N. Spangenburg (Eds.), *Dreaming planet earth*. Santa Fe, NM: Bioneers/Collective Heritage Institute. Retrieved from http://www.dreamingnewmexico.org/files/methods/view

Weiss, R. S. (1995). *Learning from strangers: The art and method of qualitative interview studies*. New York: Free Press.

Weller, M. (2002). *Delivering learning on the net: The why, what & how of online education*. Sterling, VA: Styles.

Wenger, E. (1998). *Communities of practice: Learning, meaning, and identity*. Cambridge, U.K.: Cambridge University Press.

Wenger, E., White, N., & Smith, J. D. (2009). *Digital habitats: Stewarding technology for communities*. Portland, OR: CPsquare.

Wesch, M. (2009). From knowledgable to knowledge-able: Learning in new media environments. Retrieved from http://www.academiccommons.org/2009/01/from-knowledgable-to-knowledge-able/

Wheatley, M. J. (2007). *Finding our way: Leadership for an uncertain time*. San Francisco, CA: Berrett-Koehler.

Williams, R. (1975). *Television: Technology and cultural form*. New York: Schocken Books.

Williams, R. (1980). Base and superstructure in Marxist cultural theory. *Problems in materialism and culture: Selected essays* (pp. 31–49). London, U.K.: Verso.

Wilson, C., Grizzle, A., Tuazon, R., Akyempong, K., & Cheung, C. K. (2011). *Media and information literacy: Curriculum for teachers*. Paris, France: UNESCO.

Windhoek + 10. (2001). *African charter on broadcasting*. Retrieved from http://portal.unesco.org/ci/en/files/5628/10343523830african_charter.pdf/african%2Bcharter.pdf

Wodak, R., & Meyer, M. (Eds.). (2009a). *Methods of critical discourse analysis* (2nd ed.). London, U.K.: Sage.

Wodak, R., & Meyer, M. (2009b). Critical discourse analysis: History, agenda, theory, and methodology. In R. Wodak, & M. Meyer (Eds.), *Methods of critical discourse analysis* (pp. 1–33). London: Sage.

Wu, T. (2010). *The master switch: The rise and fall of information empires*. New York: Alfred A. Knopf.

Zittrain, J. (2008). *The future of the internet and how to stop it*. New Haven CT: Yale University Press.

Zukav, G. (1979). *The dancing Wu Li masters: An overview of the new physics*. New York: Morrow.

# Index

(re)mediation 62
Action Coalition for Media Education (ACME) 41, 96, 108, 122
Adbusters 53
American technological sublime 137
anthropocentricism 29–31, 34–35, 86, 95, 103, 125, 129
Arab Spring 76, 163
Aspen Institute 22, 68
autonomous individual 37, 47–48, 57–58, 119, 158

Barber, Benjamin 81, 152
Bateson, Gregory 37–38, 46–48, 63
Bateson, Mary Catherine 150, 165
bioculture 62
Bioneers 1, 170
Blewitt, John 3, 24, 133, 139, 148
blogosphere 32
boundary object 42, 145, 149–150
Bowers, C.A. 36–37, 52, 56, 58, 61, 80, 87, 138
British Film Institute 69
Buckingham, David 1, 89–91, 93, 135, 143–144

Capra, Fritjof 9, 29, 30, 55, 133–134
Cartesian worldview 21, 37, 46–47, 120, 124, 147
Center for Contemporary Cultural Studies 69
Center for Ecoliteracy 131, 134
Center for Media Literacy (CML) 41, 96, 107
Chicago school 45, 69, 70
circuit of culture 71, 85, 145, 152, 154
citizenship, defined 28
cognition 22, 30, 47–48, 50, 52, 61–62, 70, 78, 113, 120, 122, 124–125, 130
cognitive ecosystem 27
Columbia school 69
commons 35, 53, 78, 80, 131, 166
conservationism 34
constructivism 98
Creative Commons 75, 170
critical discourse analysis 6, 97, 101–103, 128
critical theory 69, 71, 82–85, 92, 102, 150
cultural citizenship 10, 28, 67, 90, 97, 151–152
cultural commons 35, 62, 77-78, 80, 87, 91, 129, 139, 161, 166, 170–171
cultural studies 7, 23, 67, 69, 71, 82, 85, 88, 92, 98, 102, 105, 145, 150

deep ecology 34, 83, 142
deschooled society 12, 18, 162
dialogism 63
discourse, anthropocentric 35; cornucopia 35; disciplinary 5, 8, 42, 83, 98, 172; ecocentric 35; economic 2, 25, 129, 142; environmental 60–61, 82–84; environmental ideology 30, 33–34; figured world 39; industrial 35, 125; literacy 89; mechanistic 6, 31, 36, 77, 103, 119–120,

124; media literacy 4–6, 42, 83, 96–99, 101, 103, 125; postmodern 83; practitioner 105; public 29, 82
DIY/DIWO (do-it-yourself/do-it-with-others), 12–13, 14, 18, 78
double bind 38, 122
double hermeneutic 47
duality of structure 47

Earth Democracy, 34, 129, 171
ecocentricism 29–30, 34–35, 84, 86, 95, 129, 136, 171
ecocriticism 5, 8–9, 31, 36, 55, 88, 96, 101–121, 133, 136
ecofeminism 23, 34, 82, 83
ecojustice 34–35, 82
ecolinguistics 82
ecoliteracy 8, 94, 122, 125, 133–134, 136
ecological footprint 25
ecological intelligence 46, 48, 49, 52, 64–65, 79, 133
ecological mindprint 25
ecology 15, 17, 22–24, 26, 29, 31, 33, 37, 40, 45–46, 52–54, 60–62, 69, 81, 85, 122, 124, 129–130, 150–151, 158–159; definition of 32
ecology of bad ideas 46–47
ecomedia literacy 6, 17, 28, 30, 52, 61, 71, 83, 8–86, 94, 97, 124, 126, 131, 133, 135–136, 138, 143–145, 149, 151, 159, 174
Ecomedia Wheel 145, 149–152, 154, 156–158
Ecomediatone 149–150
EcoMind 49
ecosophy 82–83
ecosystem 9, 27, 32, 33, 35, 62, 65, 86, 104, 110, 112, 116, 123, 126, 129, 163
ecotone, 147–148, 150

education for sustainability (EfS) 1, 131, 133–135
edupunks 18
embodied cognition 58, 63, 150
enclosure 53, 80–81, 129, 142, 161, 170
environmental communication 16, 23, 53, 61, 82–85
environmental ecology 83
ethics 28, 30, 34, 49, 71, 83–84, 86, 123, 131, 164

figured world 39, 43, 95, 99, 101–102, 104, 115–116, 118, 120, 122, 172
film studies 59, 67, 69, 82, 105
Frankfurt school 69–70

green cultural citizenship, 1, 5–8, 28–30, 39–40, 46, 71, 75, 78, 82, 85, 87, 94, 96, 113, 122, 125–126, 133, 135–136, 156-157, 159, 161–163, 171, 173–174
greenwash 73, 76, 127, 128
Guattari, Pierre-Félix 83–84

human ecology 45, 69

Illich, Ivan 12, 65, 162
implicated actor 8, 39, 114-116
individualism 37, 77, 122, 164
Industrial and Scientific Revolutions 5, 37, 46, 70, 77, 137, 174
industrialism 35, 37, 47, 62, 137
information ecology 31, 40-41, 81, 100–101, 104, 110, 125, 139, 147–148; keystone species 41, 97–98, 101
integral ecology 145

landscape ecology 147
Lippmann, Walter 70
living systems, defined 9

# Index

Mander, Jerry 2–4, 15, 37–38, 140
Masterman, Len 67–68, 92
McLuhan, Marshall 26, 31, 45, 48, 53, 148, 161
meaning design 21–22, 59, 73–74, 118–119, 171, 174
mechanism 5, 29, 31, 35, 36–37, 46–47, 49–50, 52, 55, 57–60, 62–63, 65, 78, 84, 119–122, 124–127, 171, 174
media ecology 26, 31, 53, 60–61, 69, 71, 85, 94, 112, 120, 130, 133, 150
media ecosystem 8, 27, 30, 32, 35, 61–62, 65, 68, 76–78, 80, 100, 115–126, 129, 136, 144, 150, 163, 171
media ecotone 147, 149, 150
Media Education Foundation (MEF) 41, 96, 108, 122
media justice 96, 108, 111, 129
media literacy, critical 22, 92–93, 128, 139, 166; functionalist 22, 88–89, 106–107, 158; information literacy 22, 89–90; protectionist 23, 88, 92, 94, 108–110, 158
media literacy ecosystem 5-6, 10, 39–41, 95–99, 101, 105, 107–110, 112–114, 119, 121–122, 125, 132–133, 172
Media Literacy Project (MLP) 15, 96, 108
media mindfulness 142
media practice model 76
media studies 5, 7, 8, 16–18, 23, 32, 52, 54, 56–57, 67–70, 74, 82–88, 90, 92, 95, 98, 105, 137, 161
media studies 2.0 74, 85
mediapolis 31, 126, 170
MediaSmarts 96, 107
megamachine 46
mental ecology 83
message 16, 22, 30, 48, 56–58, 62, 93, 94, 108, 115, 118–120, 124, 126, 128, 136–137, 161, 166, 171

metaphor 5–10, 31, 37–39, 42–43, 45–46, 52, 54–55, 65, 94, 100–101, 103, 118, 125, 131, 165; analysis 103; agriculture 45, 64–65, 81, 162; conceptual 31, 97, 103, 105, 114; conduit 56; container 38, 55–56, 58, 103-105, 119, 158; conveyor-belt 56; defined 36; ecology 40, 45–46, 84–85, 144, 146, 150, 174; ecosystem 26, 32, 61, 104, 126; ecotone 147; environment 31–32, 37, 52–53, 59, 122; environmental 32, framework 36, 55; literacy 89; mechanistic 120; media 37, 42, 55, 57, 59, 60, 62, 170; message 119–120; metonyms 103; place/space 120, 125; progress 38; root 7, 36–37, 39, 52, 54, 61, 96, 125; spatial 104; syringe/magic bullet 57; systems 150; technology 40, 47; transmission 56; transportation 119–120; visual 120
mindfulness 16, 30, 142
monoculture 64, 79, 80–81, 162, 174
monocultures of the mind 78
multiliteracy 89, 143, 150, 155–156
multi-site situational analysis 6, 96–97

National Association for Media Literacy Education (NAMLE) 41, 96, 107–108, 115
Neo-Luddites 3, 24
networked fourth estate 76, 161
New London Group 21–22, 143, 150

objects-to-think-with 3, 6, 140
Occupy Wall Street 76, 81, 128, 163
*oikos* 45, 65, 139
organicism 46
Orr, David 30, 87, 173

paradigm 8, 10, 30, 35–37, 39, 74, 173; Cartesian 47; Earth 136; educational 124;

industrialism 47; mass media 85; mass society 69; mechanism 5; mechanistic 46; media 74; organic media 170
paradigm shift 30
participatory culture 75, 85
pedagogy 21, 67, 113, 127, 133, 163; critical 92, 98, 136; Earth 136; ecopedagogy 133, 136, 139; environmental 17; market-oriented 90; public 139
permaculture 64, 80–81, 162; design 145; media 65, 81
Postman, Neil 53, 112, 148
postmodernism 83
preservationism 34
Prezi 149–151, 154–156, 158
Project Look Sharp 96, 107, 115, 122, 124, 131

sacred ecology 15
semiotics 54, 69, 126, 128
Shannon-Weaver model of information 57, 70
Shiva, Vandana 34, 60, 64, 78, 129, 174
social constructionism 33, 83, 150
social ecology 34, 83
sociocultural ecosystem 27, 40
structuration 76
sustainable cultural practice 3, 6, 29–30, 58, 62, 80, 85, 95, 131, 135, 148, 165
symbolic interactionism 33
systems theory 47, 49, 60
systems thinking 1, 40, 49–50, 122, 124, 133–134, 173

*technique* 46
technoliteracy 29, 84, 136–139
thinking system 48, 62–63
Toronto school 69
transformative ideologies 34

Tyner, Kathleen 89, 91–92, 171–172

UNESCO 91, 100
unrestrained instrumentalism 34

virtual community 104

CRITICAL ISSUES
FOR LEARNING AND TEACHING

Shirley R. Steinberg & Pepi Leistyna
*General Editors*

*Minding the Media* is a book series specifically designed to address the needs of students and teachers in watching, comprehending, and using media. Books in the series use a wide range of educational settings to raise consciousness about media relations and realities and promote critical, creative alternatives to contemporary mainstream practices. *Minding the Media* seeks theoretical, technical, and practitioner perspectives as they relate to critical pedagogy and public education. Authors are invited to contribute volumes of up to 85,000 words to this series. Possible areas of interest as they connect to learning and teaching include:

- critical media literacy
- popular culture
- video games
- animation
- music
- media activism
- democratizing information systems
- using alternative media
- using the Web/internet
- interactive technologies
- blogs
- multi-media in the classroom
- media representations of race, class, gender, sexuality, disability, etc.
- media/communications studies methodologies
- semiotics
- watchdog journalism/investigative journalism
- visual culture: theater, art, photography
- radio, TV, newspapers, zines, film, documentary film, comic books
- public relations
- globalization and the media
- consumption/consumer culture
- advertising
- censorship
- audience reception

For additional information about this series or for the submission of manuscripts, please contact:
 Shirley R. Steinberg and Pepi Leistyna
 msgramsci@gmail.com | Pepi.Leistyna@umb.edu

To order other books in this series, please contact our Customer Service Department:
 (800) 770-LANG (within the U.S.)
 (212) 647-7706 (outside the U.S.)
 (212) 647-7707 FAX

Or browse online by series:
 www.peterlang.com

www.ingramcontent.com/pod-product-compliance
Ingram Content Group UK Ltd.
Pitfield, Milton Keynes, MK11 3LW, UK
UKHW022239230426
12048UKWH00018BA/1350